BLUE

PETER MATTHIESSEN, naturalist, explorer, novelist, was born in New York City in 1927 and graduated from Yale University in 1950. He also attended the Sorbonne and, in the 1950s, co-founded the *Paris Review*. He worked for three years as a commercial fisherman on the ocean haul seine crews at the eastern end of Long Island, and as a captain of a charter fishing boat. His many expeditions to the wilderness areas of the world have taken him to Alaska, the Canadian Northwest Territories, Asia, Australia, Oceania, South America, Africa, New Guinea and Nepal – memorably described in such books as *The Cloud Forest, Under the Mountain Wall, The Tree Where Man Was Born, Sand Rivers, The Snow Leopard* and *African Silences*. Peter Matthiessen has also written eloquently of the fate of the North American Indians in *Indian Country* and *In the Spirit of Crazy Horse*, and is also the author of several novels, including *At Play in the Fields of the Lord* and *Killing Mister Watson*, and of the acclaimed short story collection *On The River Styx*. Peter Matthiessen lives in Sagaponack, Long Island, New York State.

*By the same author*

## Fiction

RACE ROCK
PARTISANS
AT PLAY IN THE FIELDS OF THE LORD
FAR TORTUGA
ON THE RIVER STYX
KILLING MISTER WATSON

## Non-fiction

WILDLIFE IN AMERICA
THE CLOUD FOREST
UNDER THE MOUNTAIN WALL
THE SHOREBIRDS OF NORTH AMERICA
SAL SI PUEDES
THE WIND BIRDS
THE TREE WHERE MAN WAS BORN
THE SNOW LEOPARD
IN THE SPIRIT OF CRAZY HORSE
INDIAN COUNTRY
NINE-HEADED DRAGON RIVER
MEN'S LIVES
AFRICAN SILENCES

*Peter Matthiessen*

# BLUE MERIDIAN

## The Search for the Great White Shark

THE HARVILL PRESS
LONDON

First published in the United States 1971
by Random House, New York

First published in Great Britain 1995 by
The Harvill Press
84 Thornhill Road
London N1 1RD



A CIP catalogue record for this title is available from the
British Library

ISBN 1-86046-043-7
ISBN 1-86046-015-1-pbk

Set in Sabon by Rowland Phototypesetting Limited
Bury St Edmunds, Suffolk

Printed and bound in Great Britain by
Butler & Tanner Ltd, Frome and London

For my brother
George Carey Matthiessen

ACKNOWLEDGMENTS

To Peter A. Lake, who took all the pictures used in this new edition of *Blue Meridian*, and to all the other members of the film crew, I remain indebted for assistance, information and cooperation of many kinds, particularly to the late Peter Gimbel, whose generosity in all respects made participation in his expedition a great pleasure, and to Valerie Taylor for her kind permission to use excerpts from her expedition diary. Gimbel, Lake and Stan Waterman also made important contributions in the form of notes and letters.

Jan Moen and the crew of the *Terrier*, Captain Ben Ranford and Bruce Farley of the *Saori*, Ian Wedd of the *Sea Raider*, Captain Torgbjorn Haakestad, Reidar Smedsrud and Willy Christensen of W-29, and Captain Arvid Nordengen of W-17 were unfailingly courteous and/or hospitable; and Messrs Jim Veitch and Al Giddings submitted cheerfully to crucial interviews.

I am grateful to Dr Eugenie Clark of the University of Maryland and to Dr Roger Payne of Rockefeller University, who were kind enough to inspect the more technical material on sharks and whales, respectively; neither is in any way responsible for any errors of fact or emphasis that may remain.

I should also like to thank James F. Clark of the Museum of Comparative Zoology at Harvard for permission to paraphrase his arguments in support of the hypothesis (see pages 22–23) that the white shark's gigantic relative, *Carcharadon megalodon*, still exists.

P.M.

1971, 1975

It is to sailors the most formidable of all
the inhabitants of the sea, for in none besides
are the powers of inflicting injury so equally
combined with the eagerness to accomplish it

Jonathan Couch
*Fishes of the British Islands* (1862)

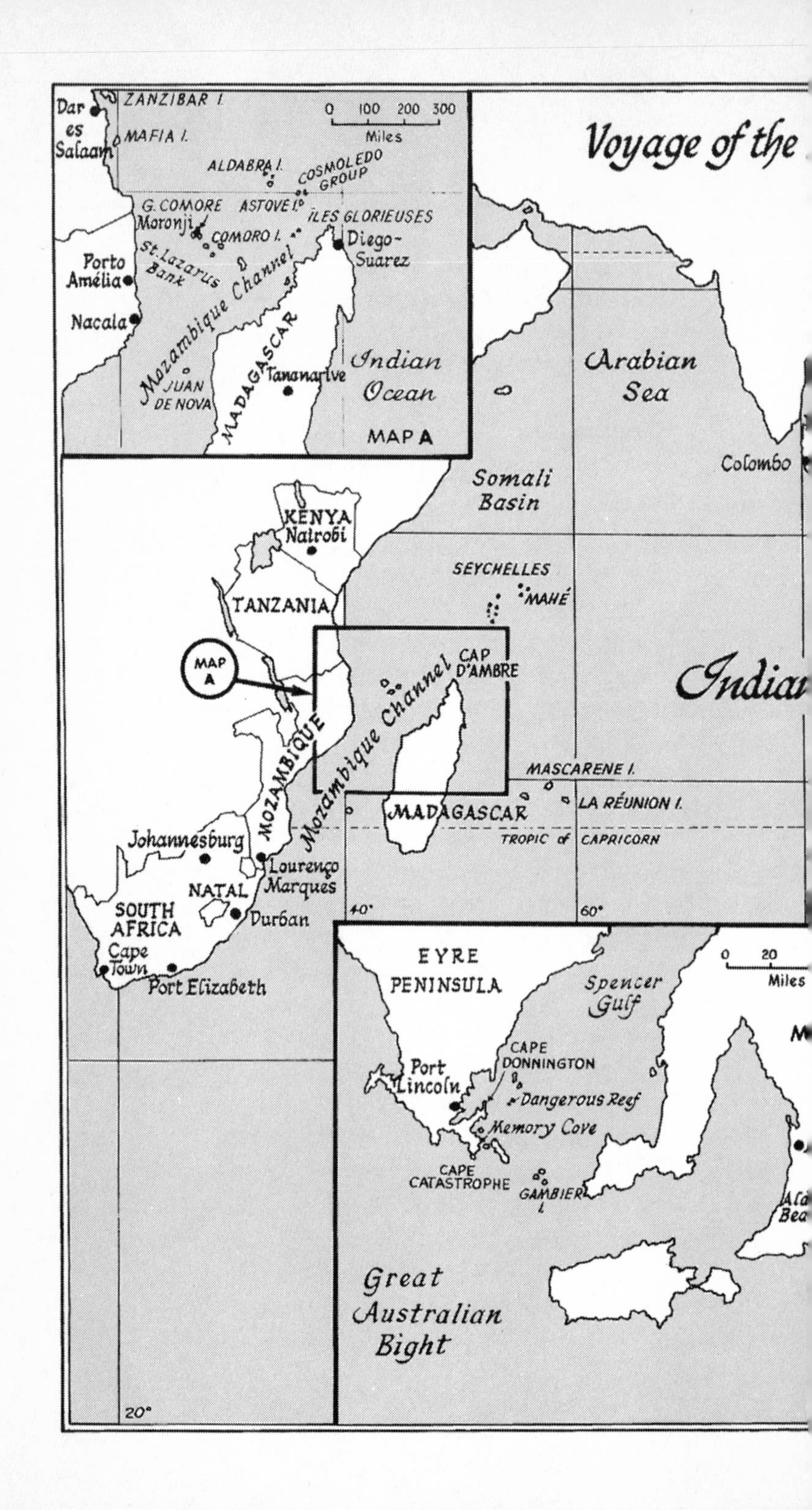
Voyage of the
ZANZIBAR I.
Dar es Salaam
MAFIA I.
0 100 200 300
Miles
ALDABRA I.
COSMOLEDO GROUP
G. COMORE
ASTOVE I.
ÎLES GLORIEUSES
Moronji
COMORO I.
Diego-Suarez
Porto Amélia
St. Lazarus Bank
Mozambique Channel
Nacala
MADAGASCAR
Indian Ocean
Tananarive
JUAN DE NOVA
MAP A
Arabian Sea
Colombo
Somali Basin
KENYA
Nairobi
SEYCHELLES
MAHÉ
TANZANIA
MAP A
CAP D'AMBRE
Indian
MOZAMBIQUE
Mozambique Channel
MASCARENE I.
LA RÉUNION I.
MADAGASCAR
TROPIC of CAPRICORN
Johannesburg
Lourenço Marques
NATAL
SOUTH AFRICA
Durban
40°
60°
Cape Town
Port Elizabeth
EYRE PENINSULA
0 20
Miles
Spencer Gulf
CAPE DONNINGTON
Port Lincoln
Dangerous Reef
Memory Cove
CAPE CATASTROPHE
GAMBIER I.
Great Australian Bight
20°

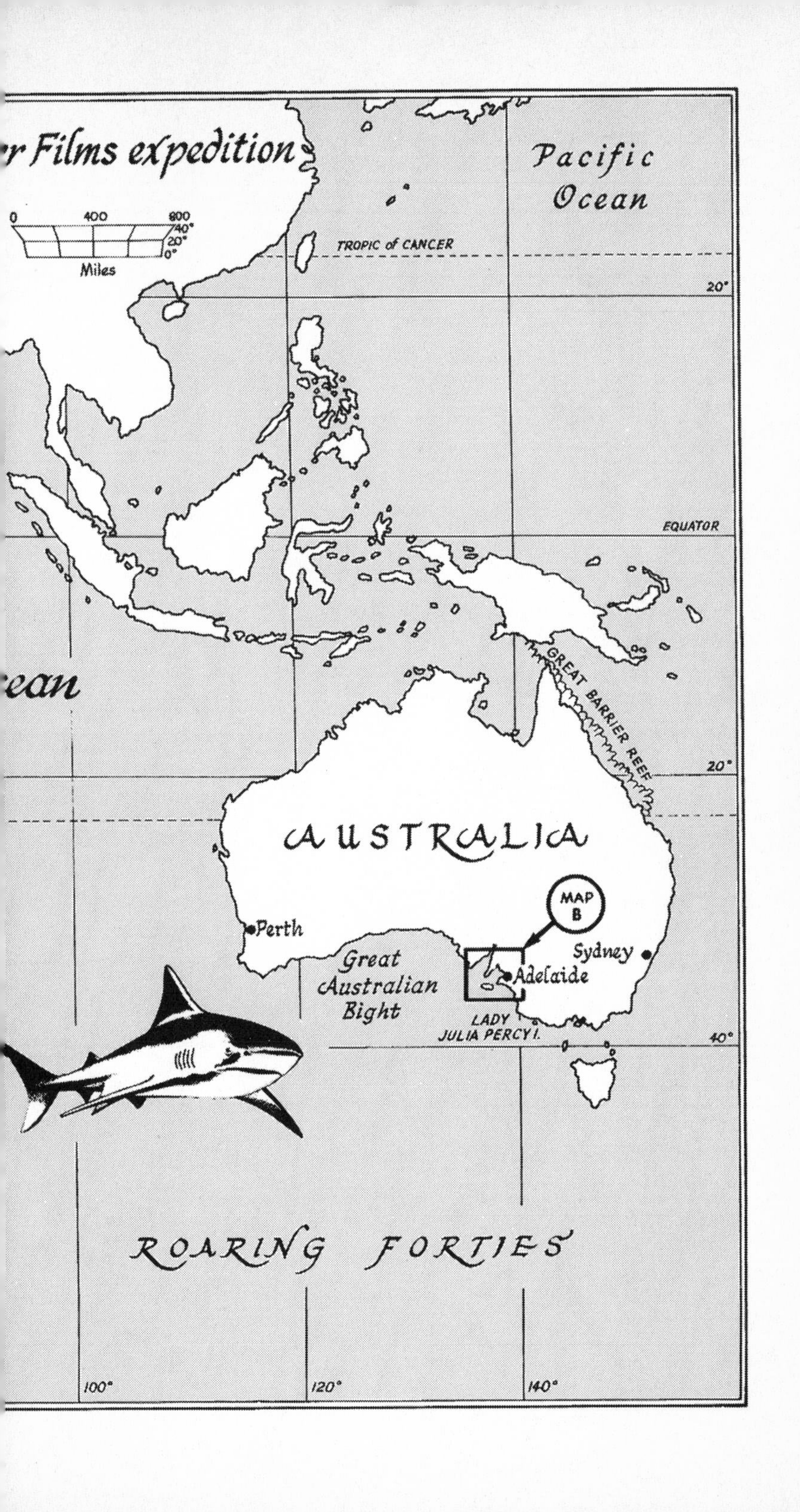
r Films expedition
0
400
800
40°
20°
0°
Miles
Pacific
Ocean
TROPIC of CANCER
20°
EQUATOR
ean
GREAT BARRIER REEF
20°
AUSTRALIA
MAP
B
Perth
Sydney
Great
Australian
Bight
Adelaide
LADY
JULIA PERCY I.
40°
ROARING FORTIES
100°
120°
140°

# BLUE MERIDIAN

GOOD FRIDAY, 1969. Aboard whale-catcher W-29, Captain Torgbjorn Haakestad, out of Durban. The coast of Natal just emerging from night shadows, twenty miles astern; no birds. Moon high over stern mast, and sun swelling the sky directly ahead, under the low cloud mass of yesterday's storms.

Storm had prevented W-29 from sailing earlier in the week. The wind whirling up the coast in squalls made it certain that the boats would stay in port on Thursday, but the office voice at the Union Whaling Company knew nothing about weather; it thought it best that any passengers should be on hand. And so I arrived at Salisbury Island docks at 3.00 on Thursday morning and sat in the guard's shack, talking to the gunsmith; he takes this shift so that he can be available in case one of the harpoon guns needs repair. In fair weather, the boats retrieve the buoyed whales at dark and haul them, sometimes a hundred miles or more, to the slipway at Durban, arriving ordinarily after midnight. At 3 a.m., after refueling and taking on water, they are bound offshore again, to be on the whaling grounds at daylight.

"Whaling is very interesting," the gunsmith said. "Very interesting." He was a short cautious man with white hair and an accent. I asked if he went often to the whaling grounds, and he said never. "I been with this same company twenty-one years and I never been out there yet." For a moment, he looked surprised himself. "I know it is interesting, but I am not curious, I guess." He shrugged. "Maybe I afraid I get seasick." Outside his window was the black harbor, and a light shine on the black blowing water. The water licked at the rusty hulls of idle whale-catchers, ranked four deep along the piers: like most shore whaling operations, this one is dying for want of whales. "Sometimes them big shark come right in here. Big ones. I see them right in here." Sharks are so numerous in Durban harbor that a shark-fishing club has been set up; its members fish from the long harbor jetties.

It was raining. The gunsmith's clock ticked on in the dead time before dawn; the air turned cold. Periodically, he announced the time, shaking his head as if, after all these years, something was imminent.

In a hard chair, the gunsmith slept – an old man in a hard chair by a black window. This was the bad light before dawn, the hour of sick dreams and thick awakenings. But when one has sat up all night, it is the hour when time stops, an hour of intense awareness; the ticking continues but the hands have stopped, and one can stand back and inspect the moment as if its mechanism were encased in glass.

The old man coughed. He had dozed like this on countless nights, hands resting mutely on his thighs, side by side like the old feet in the blunt shoes square to the floor; his feet and fingertips pointed straight at me. His stillness filled me with the consciousness of my own name, my age, the small scars and calluses on my life's hands, my solitude, my transience, and my absurdity. He sighed and so did I.

Today or tomorrow I would watch the killing of sperm whales. Later I would go beneath the sea as an observer on an expedition that had come to film the great ocean sharks as they came in to attack the rolling carcasses. Because it is senseless the whale slaughter would be ugly, but the shark's banquet would be beautiful. Such elemental life-in-death would be less horrifying than exalting, restoring the immediacy of existence. The shadow of sharks is the shadow of death, and they call forth dim ultimate fears. Yet there is something holy in their silence.

At 5 a.m., there was no sign of W-29, which was towing seven whales, nor of W-25, which was towing eight. the gunsmith kept imagining that he saw lights at the slipway across the harbor, where each whale is loaded onto a flatcar and trundled by rail a few miles down the coast to the shore whaling station. Here barefoot, bloody-legged men with flensing knives slash at it even before it is hauled clear of its car. In a bedlam of chains and machinery, great S cuts are made along its length, and chain hooks secured to the foot-thick blubber, which is hauled off in whale-length strips. The meat is stripped next, then the jaws, which are snaked away to crosscut saws to be cut into manageable pieces for the huge cooking vats buried here and there in the concrete platform.

The meat extract is essentially creatine, which is derived from the high myoglobin content in mammalian tissue; creatine is a chemical salt that awakens the taste buds, and is much used in the manufacture of soups. Myoglobin makes the flesh of sperm whales purplish and bitter, but the ton of meal derived from each five tons of meat and bone is 75 percent protein, and is used as a component of hog and chicken feeds.

Nothing is wasted but the whale itself. The head oil, or spermaceti, is used as an additive in motor oils, the teeth are saved for ivory, and the bone, blood and guts go into the manufacture of bone-meal fertilizer. The water used to wash down the great platform pours off into the sea below, where fishermen cast for whatever comes to the bloody surf.

Other whales taken are all rorqual species (*Balaenoptera*): the finback, which may reach eighty feet, the sei (Norwegian for "sail": refers to shape of dorsal fin), and the lesser rorqual, or little piked whale, known here as the minke; the sei and the minke whales are relatively small. All three are baleen whales that feed on plankton, and their meat is good – a big finback may be worth four thousand dollars, or nearly three times the value of a big sperm whale – but the catch is small and growing smaller. Blue whales, right whales, and humpbacks are disappearing from these seas, and to kill them is forbidden. It is also forbidden to kill finbacks north of 40 degrees south latitude (they breed in temperate seas), but both laws, the local whalers say, are ignored by the Japanese and Russians. In the old days, Union Whaling operated a big fleet in the Antarctic, but a few years ago it sold its factory ships to the Japanese and restricted its operations thereafter to the shore station.

The great whales of the seven seas are so depleted that Japan and Russia alone among the nations of the world are still engaged in open-ocean whaling. Though the industry now depends on the small sei whales and minkes, the remnant bands of the great whales are destroyed wherever encountered, and doubtless the lesser whales will hold out long enough to make it certain that the last of the leviathans will be exterminated along the way. Already the blue whale is practically extinct, and in most parts of the world the right whales and the humpbacks are close behind. The one large whale that still survives in any numbers is the sperm whale, which is hunted in every ocean in the world and could disappear in the next decade.

Using catcher boats and helicopters, the factory ships move ponderously about the oceans, killing ever larger numbers of ever smaller individuals. The relentless waste of life is barely profitable, since almost every whale product except ambergris is more readily available from other sources: the whaling industry is dying of consumption. Still, there is no better use for these monstrous ships and their fleets of whale-catchers – and no better use for whales, to judge from the apathy with which their slaughter has been met – and perhaps a few more years can be wrung from the investment. "Overhead" is the sole excuse for man's persistence in the destruction of the whales, and in the name of this small economy the mightiest animals that ever existed on the earth will pass from it for ever.

The whale-catchers came in just after dawn, and I went aboard W-29. Captain Haakestad had washed and was toweling his wet head. "That was a hell of a night," he said, straight off. "Fifty-mile-an-hour wind out there." Remembering my mission, he shook his head. "We not go out today. We go Friday, three a.m."

Toward midnight on Thursday, I went aboard W-29 again. There was nobody around. I found a berth and crawled into it, awaking when the ship sailed, about 3.30, and again when the first seas tried her hull; then a hand shook me. "It's ten of six!" the crewman said, accounting for his act. I got out of the berth and pitched onto the deck, where the dawn wind struck me in the face. The harpoon gun mounted in the bows, rising and falling as the boat sliced through the heavy surge, was a hard black silhouette against the sun that rose from the far reaches of the Indian Ocean.

At daylight a sailor climbs the rigging to the crow's-nest. We are thirty miles due east of the coast, headed toward a 1000-fathom depth at 14.5 knots. W-25 is off to port, and to starboard, spread out three miles apart as they fan out from Durban's Cooper Light, are W-17, W-16, W-26, W-18. The old steam engines on these whale-catchers burn crude bunker oil, twelve or thirteen tons of it each day; the engines are so simple that almost nothing can go wrong with them, and they are powerful, but in a marginal industry they burn too much fuel, and there will be no market for these ships when the whales are gone. These are the last six in the fleet; they average 160 feet in length, with a 28-foot beam. All were built after World War II in Fredrikstad, Norway, and all have white topsides on a rusty gray steel hull, with a heavy black smokestack banded in bright blue. They are rakish ships, high in the bow to insure a good angle for the harpoon gun, and so low amid ships – the work deck must be as close as possible to the water – that in moderate seas the main deck is awash; the low deck, climbing slightly once again as it sweeps astern, gives the ships a sway-backed appearance, as if they had sagged beneath the weight of their big steam engines. W-29 carries three Norwegian officers and a crew of fourteen: Captain Haakestad, First Mate Reidar Smedsrud, Wireless Operator Willy Christensen, four engineers, three firemen, a messboy, a steward, a bo's'n and four seamen.

A first bird, the sooty shearwater, slides like a shadow in the trough of the unlit sea. The wind is out of the south west at ten knots, but there are mare's-tails, and the day will freshen. In this part of the austral oceans, the prevailing winds are southwest or northeast; a rare and violent storm wind that lashes and carves the open coast is known as a black southeaster.

A radio call from W-25: she has sighted whales. Although the quarry is ten miles away, the captain abandons the big spread of eggs and bacon, smoked herring, beans, chili peppers, brown bread and milk; he is catch leader of the fleet, responsible for the position of the boats.

Northeastward. The whales are fifty-six miles at sea, on a 78-degree bearing from Cooper Light. The spotter plane, turning in high silence in the ocean distance, has located two pods several miles apart, headed slowly south.

A giant bird, bone-white but for upper wing coverts and under wing tips, and a thin band at tail tip and wing's trailing edge, all of these black; its beak and legs a pale pink, like a sun-worn conch shell high above the tide line – this is the wandering albatross of the southern oceans, the greatest flying bird on earth.

At 8.15, W-25 is broadside on the horizon, blunt black on a silver sun. She is killing whales. Slowly she turns in the pall of her own smoke and is under way again. Another whale blows near the yellow-and-green flag that marks W-25's first kill; in the morning wind the bright flag snaps on a bamboo pole that rises twelve feet from the buoy float. Fixed to the pole is a small radio transmitter so that the buoyed whale may be located from a distance, even after dark. In the old days, radios were unnecessary: the ships did not have to wander far to find another whale. Even four years ago, the season's catch of sperm whale out of Durban was three thousand animals; last year it was eleven hundred. This year it will be even less, and next year, so it is said, the shore station will shut down for want of fodder.

At W-29's approach, the whale has sounded. Willy Christensen leaves the boat's high bridge to take up his post in the radio shack, and a few minutes later his voice rises eerily from the tube: he has located the whale with his machines. The captain grunts something at the helmsman, who alters course and signals to the engine room to halve the speed. Captains prefer to be called gunners and are hired primarily for that talent whether or not they have a master's ticket. Now the Gunner leaves his seat in the starboard corner of the bridge and walks the long sloping catwalk that leads from the bridge over the litter of chains and hawsers and big winches of the foredeck to the bow. The gun on the bow platform is tilted so that the heavy barb on the harpoon is pointed downward at the water, and flyingfish skid outward, veering away between the waves.

The whale remains at a thousand feet. On the bow, the Gunner sits on the catwalk step, big hands resting on his knees. He is a stolid unpretentious man with a gentle voice, and he is patient. The sonar can track the whale a half mile down, unless strong underwater-current lines intervene. "With us they stand no chance," the mate says quietly, a trace of weariness in his tones, perhaps regret.

Christensen's voice rises again, and the mate gives the helmsman a new course. The boat slows to a hum and glides forward in strange stillness, the slosh of water audible along the hull. Below, the whale rushes through the dark, driven by the relentless *ping* that it is unable to escape, and above, the steel boat waits, rolling heavily in the monotony of seas. The sperm whale can submerge for an hour or more, but this time is rapidly decreased by panic when the animal is pursued. Peering down into the silent sea, I wondered what sort of awareness tuned the minds of those great hellish shapes so far beneath. The sperm whale has a bigger brain than any animal that has ever evolved, and unlike the baleen whales, it is not a grazer but a hunter; other cetacean hunters such as the porpoises are very intelligent animals indeed. Doubtless these doomed creatures were communicating their alarm, though whether they have the vocal range of the baleen whales is doubtful: a herd of humpbacks (and probably other baleen species as well) sing like the horns of paradise, arriving at harmonics not attainable by the instruments of man. For the sperm whale, only *clicks* are known – sometimes these sounds are audible during the last throes of the harpooned whale – and it is possible although not likely that, like the humpbacks' low-frequency notes, these *clicks* can carry enormous distances and are designed to do so. (It is now believed, from preliminary evidence, that the deepest and most sonorous notes of a humpback whale can and may be heard by another humpback anywhere in the same ocean basin, and may even resound around the world. Cosmic sounds, electronic sounds, the music of the spheres shimmer through the soft gurgle of the sea with the resonance of an echo chamber, and with them soft bell notes and sweet bat squeaks, froggish bass notes, barks, grunts, whistles, oinks, and elephantine rumblings, as if the ocean floor had fallen in. No word conveys the eeriness of whale song, tuned by the ages to a purity beyond refining, a sound that man should hear each morning to remind him of the morning of the world.)

A patch of bright-brown sargassum weed. A storm petrel. W-29 rolls and wallows, and the man in the crow's-nest swings in crazy arcs on the morning sky.

The whale has vanished into the abyss. Sperm whales can descend to at least seven thousand feet (a two-and-a-half-mile round trip from the surface), and this one has escaped the sonar. The radioman, reappearing, shrugs. W-25, tracking another whale, wheels across our bows, full speed ahead, heeling over to near 45 degrees as she makes her turn. The sun glints on the sky-blue band of her black stack, amidships: on the band is a big *U* for Union Whaling Company.

9.10. The helmsman, a dark figure with patches of fair skin on neck and throat, points to the eastward: a wash of white water is subsiding where two whales have broached. But these whales, sounding, are assigned to other boats that come up rapidly from the south. W-29 is bound offshore, where the plane, like a black hornet on the sun, is circling.

The mulatto helmsman with white throat patches is classified "white", whereas two of his mates who would be white men elsewhere in the world are classified in this rigid land as "colored". Their officers are perplexed by apartheid but not offended; it is not their business. Like all South Africans, they ask uneasily how one likes South Africa, and I answer by not answering. They nod, still perplexed, still not offended. After all, their expressions say, we are Norwegian. Later I asked Willy Christensen how *he* liked his adopted land, and he sighed. "Nothing ever happens here," he said. "But I been here so long, I don't know nothing better any more."

A solitary tern, sixty miles offshore, is gone again before I catch it with binoculars; the boat rises on a swell, and when it descends, the swift white bird has vanished like a wisp of spray into the infinities of sea.

Where the pod was first sighted, the plane has dropped a dye marker, and an unnatural bright stain of plastic green rises and falls on the cruel blue. A mile further, the ship comes upon the whales. Fleeing bad vibrations, they are headed rapidly offshore, porpoising strongly through the choppy seas. There are more than twenty in formation, and the family groups remain tight together, swimming abreast. On every rise, the sea pours off the glistening black backs; then the mist of their breathing disappears in an explosion of white water. Clouds cross the sun, but the surface of the sea, still reflecting light, is a strange dead silver. The sun returns. The wind is rising, and when the whales blow, a rainbow appears in the fine mist as it drifts downwind.

9.48. On the bow, the Gunner's heavy form rises and falls, breaking the line of the horizon. The harpoon gun swings from port to starboard, searching for the biggest whale, but these close groups are composed of cows under thirty-three feet; he finds no target. The bow cuts the pod in half, and the whale shapes, fleeing to the side, slide like cloud shadows beneath the sea. In moments, they rise, surging and blowing, and the ship rides down on them. The Gunner raises one hand almost casually to point; he bends to his gun again as the ship surges. A loud thump on the wind, muffled and ear-stunning – the mate runs forward along the catwalk to reload the gun.

Oddly, the shot has missed; the Gunner sits down heavily as the mate reloads. A sailor lugs a slotted red harpoon from the foredeck and with the help of the mate jams it into the muzzle. Inside the four flights of the point, the harpoon, four feet long, weighs 185 pounds. The mate sets the explosive grenade that detonates in the whale's body three seconds after impact, and with a few turns of light lanyard secures the harpoon against sliding out of the tilted muzzle. Then the human silhouettes retire from the bow, all but the form of the seated Gunner, as black as the gun itself on the sparkling sea. Less than three minutes have elapsed, and already W-29 is circling in on another pod of whales, so close that one can see the distinct forward angle of the sperm-whale spout that issues from a hole just to the left of the center line of the head. (The Gunner says that this hole is closed when a sea washes over it and that sometimes a whale in flight will try to spout underwater.) Ploughing and blowing, the whales leave a white wake in the blue, backs gleaming like smooth boulders of obsidian in a swift torrent; as W-29 comes down on them, they sound. One black back arches into a curve, and a huge fluke rises in slow motion from the sparkling ocean. Water cascading from its fluke, the whale slides down in silence into the sea.

At 10.15, as the ship heels into position, the whole pod broaches in one mighty burst of mist and spray. The ship rides herd on the black backs, the harpoon point still seeking a big whale. At 10.19, a series of explosions: the shot, the muffled boom in the whale's body, and the jolt of a huge spring belowdecks. The nylon harpoon line, with a breaking strength of twenty-four long tons, is reeved through chocks under the gun platform, then up over a heavy pulley under the crow's-nest, high on the mainmast, then down again to heavy winches at the aft end of the foredeck, under the superstructure; the pulley is rigged to enormous springs under the foredeck that take the main impact of the whale's first thrash.

The whole ship quakes. The dying whale has veered away to starboard, and the harpoon line shivers spray as it snaps taut; the white of the cachalot's toothed lower mandible flashes in the light as the beast rolls, and the first well of its blood spreads on the surface. With her winch, the ship is warped alongside, and the mate puts a killer harpoon – a grenade-carrier with no line attached, known to the Norwegians as the flea – into the thrashing hulk as the crewmen jump to dodge the wall of spray. Now the whale is still; only the pectorals twitch a little as the last life ebbs out of her. Already a long pole has been used to jam a hose tube into the carcass, and air is pumped in to make sure the whale will float. At the same time, the sailors rig a heavy noose around the base of the fluke, which in turn is secured to the big float of the marker buoy. With a flensing knife lashed to the end of a long pole, the mate, doubled up over the gunwale, cuts the harpoon line where it is spliced to the imbedded missiles, and the ship backs off from the buoyed whale. The harpoon gun has already been reloaded, and a new line spliced to the harpoon. Eleven minutes have passed from the moment the first iron struck the life out of the whale until the whale-catcher reverses her screws and backs away.

The inflated whale lies on her side, washed by red waves of her own blood. Already the bright stain on the bright sea is huge and thick, as if it would never wash away. The blood spurting from the wounds is a deep mammalian red, but on the surface of the sea it turns red red, as vivid as a dye, and the amount of it is awful.

By international convention, not observed by the Russians, these men say (they very much resent it that the Russians and Japanese, having copied all the Norwegian techniques, now dominate the whaling industry), a sperm whale less than thirty-five feet in length – thirty-eight feet in the Antarctic – may not be taken, although a two-foot leeway is granted in acknowledgment of the difficulties involved. A gunner's fine of thirty pounds that is levied by the government on each whale under thirty-three feet is considered part of the overhead. In whaling terminology, a small sperm whale runs from thirty-three to thirty-eight feet, a medium one from thirty-eight to forty-three, and a big whale is anything larger. This whale looked undersized to me, and I asked Willy Christensen what he thought. He grinned, jerking his head toward Haakestad, who was rolling slowly up the catwalk. "Better ask the Gunner," he said.

Before returning to the mainland, the spotter plane reported that seventy whales were in this area, and for a time there were spouts everywhere, but the other boats were also having difficulty in locating whales of legal size, and W-29 steamed further offshore to look for big bulls that might be attending the main pod. For three hours the ship searched hard, with no success. Often, in the middle of the day, the whales seem to disappear, and it is supposed that they are hunting giant squid in the ocean depths, coming up infrequently to breathe. The wind had risen to a 25-knot blow, and spouts would be hard to see in the whitecapped water; dark shearwaters arched across the iron bows like boomerangs and, beyond, the world was empty.

Noon on the meridional seas. Blue sky, blue sea, a ray of sun reflected from the deeps like a blue meridian.

By early afternoon the six boats had harpooned but six whales between them, and since W-29 was short two men on its crew, Captain Haakestad decided to tow the six whales into port. The other boats would hunt all afternoon and if weather permitted lay to at sea until next morning.

Though the whales had been floating for several hours, two of them had drawn no sharks at all. The rest were skirted by a few oceanic white-tip sharks, tawny ochre in color: they slid in and out of the red roil and wash of whales and hull, unhurried, almost motionless. Each whale had one albatross or more, picking red gobbets from the blood pool, and one was courted by a pair of storm petrels, dancing and fluttering like black butterflies.

Below, behind, and further out, in the blue shadows, other sharks hung back at the ships' approach; there was no sign of the frenzy that in the past ten days had stripped five buoyed whales to ragged backbone. According to the spotter plane, one of the stripped whales was a large bull; its forty to forty-five tons of flesh were gone in half an hour. The Norwegians said that twenty to thirty big sharks raging at a whale carcass was not an uncommon sight, and at these times the sharks were not deterred by ships; sometimes they surged halfway out of the water, or lay on the carcass for fifteen seconds at a time, gobbling at the flesh above the surface.

The ocean off South Natal is notorious for its sharks; there are two and a half times as many shark attacks off the Durban–Port Elizabeth stretch of coast than in all the rest of the east coast of Africa, from Cape Town to the Gulf of Aden, and the offshore waters of Madagascar. On November 28, 1942, the British troopship *Nova Scotia*, carrying Italian prisoners of war, was torpedoed on a bright calm morning in these waters, and a minimum of seven hundred people died, many of them victims of a shark attack that went on for hours. The few survivors owed their lives to the U-boat captain, who surfaced, perceived his mistake, said, "I'm terribly sorry," saved two of the swimmers, and risked his own security by transmitting a radio call for help.

Each whale was detached from its buoy, then winched up by the bridle on its fluke to the port side. When the base of the tail was lashed tight to the hull by a heavy chafing gear of rope-wrapped chain run through a hawsepipe, the enormous flukes, surging above the level of the rail, were trimmed and notched by the mate's flensing knife in a way that indicated which ship had caught which whale; when he sliced the tips off the heaving flukes, the shining black wings skittered away into the sea.

The first whale was winched tight to the hawsepipe furthest aft, and the next ones were secured one hawsepipe forward, in succession, so that whales already taken in tow would lie alongside as the boat eased along and not impede the operation. There are six hawsepipes on each side of the ship, but all whales were secured on the port side; the ship lay broadside to the wind on her starboard beam so that the work could be accomplished in the lee.

In the late afternoon, W-29 collected her last whale and headed west on the long voyage to Durban. Heeling to port, she was slowed perhaps three knots by the dragged carcasses, which writhed and twisted in the wash and sea surge as if come to life. Crashing together, their graceful flukes cropped, jaws slack, tongues and guts protruding, they looked damned, and the infernal atmosphere as twilight came was not lessened by the clank of the rough chains that sawed the hides, nor the din of the sea's rush against the carcasses, nor the rank wake of mingled blood and feces that washed out of them into the darkening blue turquoise of the wake. During the harvest of the whales, the ship had collected the attendant albatrosses, and the giant birds wheeled up and down, wings motionless, slooping now and then to pluck a scrap, then climbing again until the wind seized them. Far astern they wheeled and fell like whitecaps blown free from the hard afternoon sea, and the pale shadows were still visible at darkness.

At supper, the Norwegians plied me with hot peppers, which on this ship are consumed with every meal; they laughed heartily at my tears. We discussed the expedition of cameramen-divers that was here in Durban to film sharks on the whaling grounds, including, if possible, the great white shark, an aggressive man-eater known locally as the blue pointer. The divers would descend in cages into the murk around the whale; the cages were light, with flotation devices to control their vertical motion, and though the water here was one mile deep, they would not be attached to the ship. The whalemen whistled. Learning that I would go down in the shark cage, they shook their heads. "Here," somebody said, shoving the pickle jar across the mess table, "better have some more hot peppers."

I lost a chess game to the Gunner and turned in. When I awoke, some time after midnight, W-29 was off the whale slipway in Durban. Under the lights, the black whales glistened in the murk, which looked thick as petroleum. I thought of the gunsmith, and of the harbor sharks following the blood trails through the night below.

Saying goodbye, Captain Haakestad said that in all his long years with the whaling company he had never bothered to pay a visit to the factory, which is only ten minutes from the slipway. Like the gunsmith, he spoke shyly, in surprise, as if vaguely troubled by his own lack of curiosity. Probably he would never go. It was said that in 1970 this shore station would close down, and he and the mate were starting a shop in the suburbs and a life of retail merchandising. He had never liked the sea, the Gunner said, and would not miss it.

IN NEW YORK, in 1968, Peter Gimbel had talked about his Blue Water Films expedition, which would hunt large sea creatures at several remote locations in the Indian Ocean, including the whaling grounds off the South African coast. Since I planned to be in East Africa in any case, and had never seen a whaling operation, much less big sharks at close quarters underwater, I accepted his invitation to come along as an observer. How would his film differ from the Cousteau shark films that were currently appearing on television? "I said 'big sharks'," Gimbel said. "Those films are fine, but I didn't see a shark over nine feet. The sharks we want will be dangerous to divers, and the shark we want most is the great white shark – in fact, the film is a search for the great white shark, whether we find him or not." The great white shark, which may exceed a length of twenty feet, is much the most dangerous creature in the sea.

Since I had no real experience underwater, Gimbel suggested that I come that summer to the Bahamas, where he would be running tests on equipment and crew. There I could make a descent or two in his shark cage, and decide whether I still wished to try it in the Indian Ocean.

In late July, I joined the film crew at Hog Island, across the channel from Nassau, where a house had been rented that had a dock and even a small workshop-laboratory. I was fresh from my first two diving lessons in the Florida Keys, where neither instructor had accompanied me into the ocean, and where I had been forced to go to the aid of a fellow student whose straits were scarcely more dire than my own. My third lesson, which came from Gimbel, was more helpful than the other two put together.

After twenty years of experience, Peter Gimbel is one of the best divers in the business; it is he who obtained for *Life* the first pictures of the *Andrea Doria* lying in 225 feet of water in the treacherous, dark North Atlantic currents off Nantucket. He is also a first-rate teacher, taking the trouble to explain the theory of diving as well as the practice of it; he knows that an extra scrap of knowledge might save your life. More important still, he dives with you and watches you and sets up such small tests and emergencies as pulling out the mouthpiece of one's air line; these things can happen by accident underwater and may cause an inexperienced diver to panic. (It is panic, not true danger, that kills most divers. Drowning is a far more common cause of death than the celebrated air embolism, which itself is ordinarily a result of panic: a frightened diver decides he must get to the surface *now*, and grabs a big breath of air which he forgets to expel as he ascends. Near the surface, where ambient pressure is rapidly decreasing, the pressure of the air still in his lungs is no longer equalized, and his big breath, expanding, ruptures his lungs.)

Finally, Gimbel inspires trust. He knows his business and he knows what fear is – one should never entrust one's life to a fearless man – and in a crisis he is steady; he lets the fear show later, when it doesn't matter. In short, he is a professional; if he weren't, I would avoid his shark cage, which adds claustrophobia to the fear of suffocation that causes the beginner, gasping for breath, to burn up so much air. After my first dives with him I had some humbling evidence of my own apprehension and inefficiency; there was half again as much air left in Gimbel's tanks as there was in mine.

I stayed long enough in the Bahamas to make three descents in the shark cage, and on the last one we went down into the Blue Hole, a huge black well in the ocean floor out eastward of Rose Island, on the Yellow Banks between New Providence and the Exumas. The mouth of the Blue Hole, two hundred feet in diameter, lies forty feet below the surface, and when both sky and sea are clear and the wind moderate, it is plainly visible from a boat's deck, not only because it looks perfectly round – an unnatural and sinister shape on a sandy bottom – but because the black of it is so much blacker than the cloud shadows that drift across the banks or the scattered beds of turtle grass and coral. The Hole drops 180 feet straight down into the darkness, and at the bottom are two passages that must bring in deep water from the Tongue of the

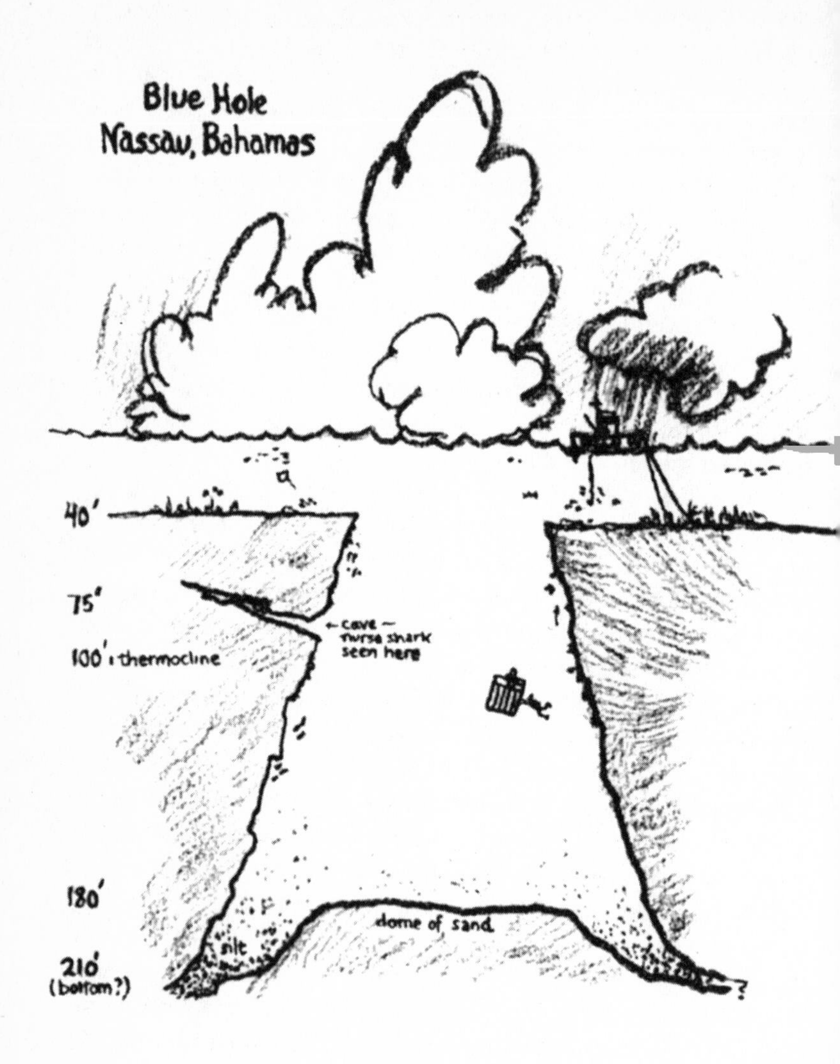
Blue Hole
Nassau, Bahamas
40'
75'
100' : thermocline
← cave —
nurse shark
seen here
180'
dome of sand
silt
210'
(bottom?)
?

Ocean, twenty miles away, because the lower part of the Blue Hole is very cold indeed.

The shark cage is not mobile horizontally, and the currents carried it thirty feet or more beyond the south rim of the Hole. By injecting compressed air into the flotation chamber in the center of the roof, Gimbel adjusted its buoyancy so that it hovered just above the sand; then we left the cage and wrestled it through the water. The aluminum cage, six by six by three and a half feet, looks delicate and is, but its air tanks, steel fittings and 110 pounds of lead ballast make it unwieldy underwater. The silent struggle was attended by big graceful African pompano with long pale dorsal streamers, flirting and turning at the abyss cage. Then the cage drifted over the rim, magically suspended above the void: with its batteries, pneumatic solenoid valves, amplifiers and power switches all hidden in the choke housing of the flotation chamber, the cage looked weightless, as mysterious in its source of power as a spaceship.

Just below the rim of the Blue Hole were rock gardens of sessile sea life – algae, corals, anemones, hydroids, and the flower-like filaments of fan worms – but I took small pleasure in them. We entered the cage, and I swung the door to behind us; there came a distant underwater click as I shot the bolt. At the control panel Gimbel made his customary sign; he lifted thumb and forefinger together and raised his eyebrows – Everything okay? – behind his mask. When the sign was returned, he pushed upward on the vented pipe that controls the entry and exit of air in the flotation tank over our heads. Escaping through the vent into the sea, the air made a cavernous booming, a sound of doom, or so it seemed to me, and the fragile craft sank swiftly through the sea floor, down the looming cliff face into the darkness.

Watching the needle on the depth gauge and straining to clear my mask and ears, I did my best not to gasp all my precious air away. Over our heads the round mouth of the Blue Hole was taking shape against the sunlit surface of the sea; I felt immersed in a heavy element, like a fly in amber – the surface above appeared liquid.

Abruptly, at seventy feet, the cage was locked in a vault of dark cold water. In the Blue Hole, the thermocline is so well defined that one's legs may be in the frigid abyss and one's chest in the tropic sea of the

Yellow Banks: Gimbel checked the cage at just this point to prove it. Then, shaking with cold, we continued the descent. Far, far above, the black outline of the rim grew smaller, and the silver ripple of the surface, farther still, was a remote gleam of hope; it was the loveliest and longest 110-foot view I have ever had.

Gimbel pulled down on the control pipe, bringing a sensor into contact with the sea water; water is a conductor, and a circuit was made that opened valves connecting one of the twin tanks of compressed air to the flotation chamber. Miraculously, after a shudder, the cage drifted upward toward the light. In an access of joy, I extended my hand – it had a white canvas work glove on it, I recall, and looked rather disembodied – and it is one of the things I like about Gimbel that he understood this impulse that could not be verbally expressed. Like two mad inventors who are testing a weapon of doom, we shook hands solemnly beneath the sea.

On the ascent we left the cage suspended at eighty-five feet and swam out to explore a cave in the south wall of the Blue Hole. The mouth of the cave, which goes back an unknown distance, lies beneath an overhang, and the opening itself is so narrow that one must enter it single file. Gimbel let me go first, the better to see the cave, which he illuminated over my shoulder with a strong underwater beam. My weights were too light, and my tank kept clanging on the cavern roof as I pulled myself forward, and I could not help remembering a story told by a friend who once got stuck for fifteen minutes while exploring a passage like this one in the south cliffs of Grand Cayman. He could go neither forward nor back, and the passage was too tight to shed his harness; also he was all alone, which divers should never be, and he was low on air. He used up some of the remaining air controlling his panic, which would certainly have killed him, and at last contrived to slip his tank and break for the surface.

Ahead of me, on a ledge, I was surprised to see a bright-red squirrelfish – had we driven it into the cave or did it live back here, in the pitch darkness? – but not nearly so surprised as I was a moment later, when something gave a sharp tug at my leg. Seeing nothing behind but the bright eye of light, I thought I had kicked Peter in the face; I waved my hand by way of saying sorry, and started forward again. This time the tug was violent. The light was switching back and forth, its beam on the gray limestone wall, perhaps twelve feet ahead. The whole wall was revolving, and it had a dorsal fin, and as it continued turning I saw the caudal fin of a large shark that we had trapped by accident in the cavern. Already the light was retreating to make room for me, and I backed up clumsily, eyes on stalks, like a crippled lobster.

Awaiting us in the open water, evanescent as a jellyfish in the blue sunrays, the empty cage had the geometric transparency of a diatom magnified a million times.

Our air almost gone, we went to the surface. The shark in the cave had been turning inward in a tight circle, and as I had not lingered there to inspect its head, I was not sure what species it had been. "A nurse shark," Gimbel told the men on deck. "Eleven or twelve feet – the biggest I've ever seen." Ordinarily the nurse shark is harmless, but this one, driven deep into the cavern, had been trapped in a small space where it could scarcely turn around. There was much less danger that it would attack than that it would knock us senseless against the coral or cut us up with the placoid scales of its rough hide in a charge to freedom through the narrow passage.

Peter was pleased by my passivity throughout the dive, putting the most generous interpretation on what was probably some weird kind of catatonia. Perhaps I had burned up all my dread on the descent down that darkening shaft into the unknown, because nothing that happened on the way toward the sun could possibly have dampened my euphoria.

"You're ready," he said with his usual optimism. "I'll dive with you anywhere and any time. But when we have a shark frenzy around those whales, the cameramen come first, you know; I can't promise that I'll get you in that cage." I told him not to go to extra trouble over it; I was feeling an ambivalence eight months ahead. Before leaving the Bahamas, I happened to say that to judge from the damage I had seen sharks do to commercial fishing nets, which had plenty of give in them, a big shark that meant business would part that aluminum cage like a bead curtain. Gimbel was annoyed that I said this in front of the others but he did not deny it. Still, I dreaded the sharks much less than drifting down a mile or more in a disabled cage in the darkness of the Indian

Ocean, unable to swim upward through the milling sharks, yet unable to stay.

"It's a funny thing," Gimbel remarked in Nassau. "A lot of divers, even very good ones with plenty of experience, are spooked by that damned cage – I don't understand it." But he himself would have every right to be spooked by his own invention, which very nearly cost him his life.

In the summer of 1960, Gimbel had come upon a crude old cage in the backyard of Captain Frank Mundus, whose charter boat, the *Cricket II*, sails out of Montauk, New York. Captain Mundus specializes in sport fishing for sharks, and the cage belonged to a client who was curious to see what was going on around the baits.

The Montauk device was essentially a steel lion cage that could be lowered on a pulley. Lacking any control of its natural tendency to head straight for the bottom, it remained tethered to the boat like a sort of sea anchor, yanking up and down with the boat's motion. This was physically disagreeable – Gimbel discovered that one can get seasick underwater – and no help at all in underwater filming, though some still shots he obtained from this contraption were reproduced by the *National Geographic*.

In the years that followed, while studying mathematics and physics at Columbia and leading an expedition to Peru, Gimbel thought about a self-sufficient cage, and in the fall and winter of 1964–65, with aid in electronics from an engineer named Mitchell Bogdanowicz, he built a prototypic apparatus. The thing had a bottom constructed of chicken wire, and depended for buoyancy on two stainless-steel beer kegs that had to be manipulated separately, but the electronics and the pipe-vent system were essentially similar to those in the present model.

Philip Clarkson, who was a witness to its maiden voyage in the spring of 1965, "never thought much of that cage." Clarkson had met Gimbel a few months earlier in San Diego, where both were working on a film about gray whales with Bob Young, the cameraman-director; Clarkson, who owns a tree nursery in New Jersey, had plenty of free time in the summers, and had served Young as production manager for Young's feature film *Nothing But a Man*. Gimbel asked Clarkson to manage a proposed film on sharks to be shot that summer off Montauk, and one day in May, Clarkson accompanied Gimbel and marine biologist Carleton Ray to a water-filled quarry in White Plains, New York, to observe the maiden voyage of the cage.

In due course, Gimbel's invention was wrestled into the water, where it was struck with a bottle of champagne. Then, with Gimbel and Dr Ray aboard, it disappeared from view. Almost immediately it got cocked on a ledge and toppled over, dumping its air, whereupon – still on its side – it continued its journey to the bottom, seventy-five feet below. Ray and Gimbel made good their escape, and the cage was rescued later. Clearly, what it needed most was a stabilizing system, and Ray suggested the permanent flotation tanks that were later installed on the four sides of the upper rim. The tanks not only gave positive buoyancy – the cage, that is, would go to the surface without mechanical assistance as soon as lead ballast inset in the cage floor had been released – but insured that the cage would remain upright under all conditions.

That summer of 1965, the new cage was used for *In the World of Sharks*, a fine short documentary on blue sharks shot off Montauk from the *Cricket II*. Gimbel and Young served as cameramen, and Clarkson as production manager. A year later Gimbel tried to film the *Andrea Doria* from the cage, but after ten days of wind, fog, dark ocean depths and evil currents, the threat of collision in the steamship lane, leaking camera housings, an overturned skiff, and other mishaps, this adventure was abandoned. The following winter, negotiations with a CBS subsidiary called Cinema Center Films for a feature documentary on sharks encouraged Gimbel to embark on a year of research that concluded, in 1968, with a location trip to Africa and Ceylon, followed by the summer of testing of equipment and personnel at Nassau. By this time the kegs had been replaced by a centrally located control chamber incorporating a cylindrical choke that insured precise control of the volume of the air bubble, thereby reducing rough vertical oscillation to a minimum; these precise controls permitted the cage to be left suspended at any depth while its occupants explored the open water.

Gimbel's concept of a feature film was already developing when he made *In the World of Sharks*. In June 1964, the *Cricket II* had caught a great white shark seventeen and a half feet long that weighed over four thousand pounds, or the equivalent of twenty-five big men. This monster was towed ashore and hauled out on the docks at Montauk, and as I lived not far away I went down one day to see it. Though white sharks vary in color, or lack of it – some are gray-white or brown or even a bruised blue – they tend to pale as they grow older, and this one was a dirty grayish white, like a cadaver. Its length was awesome, and so was its vast maw, but most appalling was its girth, its massiveness: one saw immediately how such a beast could take a seal in a single dreadful gulp.

In nineteenth-century Samoa the white shark was highly thought of, due to its habit of devouring thieves who stole fruit from the people's trees, but elsewhere in the world it had acquired a very evil reputation. Thomas Pennant, in 1776, noted that it grew "to a very great bulk. Gillius says that in the belly of one was found a human corpse entire, which is far from incredible, considering their vast greediness after human flesh. They are the dread of the sailors in all hot climates, where they constantly attend the ships in expectation of what may drop overboard. A man that has this misfortune perishes without redemption. They have been seen to dart at him like gudgeons to a worm . . . Swimmers very often perish by them. Sometimes they lost an arm or a leg and sometimes were bit quite asunder, serving but for two morsels for this ravenous animal."

Because of anatomical similarities, notably the symmetrical crescent caudal fin (most sharks have asymmetrical tails, with the upper lobe much larger than the lower), this species is classed with the mako and porbeagle sharks in the family Isuridae, the mackerel sharks, but a closer relative is *Carcharodon megalodon*, which became extinct in recent times, perhaps no earlier than ten to fifty thousand years ago, and exceeded a length of eighty feet: the extinct form and the modern one are assigned to the same genus because they share the triangular serrate teeth. According to shark literature, C. *carcharias*, the great white shark, may grow to thirty feet or more; the largest specimen of recent years, found tangled in heavy chain off Port Fairy, Australia, was measured at thirty-six and a half feet overall. The many reported sightings of enormous white sharks have led a few ichthyologists to wonder if C. *megalodon* might still exist. It has even been argued (see Acknowledgments page) that fossil teeth of small *C. megalodon* are so similar to those of living *C. carcharias* that the two may represent a single species. In any case, it seems unlikely that this once cosmopolitan and abundant shark but not its near relative should have disappeared for lack of food, which is the usual cause given for its extinction. A shark will eat virtually anything and can feed sporadically when necessary, storing nourishment in a huge liver that at times comprises one quarter of its body weight. The oceans are still supporting whales, which have a higher metabolic rate than sharks, and possibly *C. megalodon*, like the sperm whale, feeds in the ocean depths on squid, which have been found in the stomach of the white. A giant shark would not need to surface as whales must do, nor scavenge on the continental shelf, where the smaller whites are mostly seen. The absence of an air bladder in sharks precludes the possibility of meaningful detection in the deeps by sonar, as well as the likelihood that a dead specimen would drift ashore.

In 1918, off Port Stephens on the east coast of Australia, a great pale shark was reported that gulped down several crayfish traps, three and a half feet in diameter, "pots, mooring lines, and all." The estimates of its size by those who saw it seemed "absurd" to Dr D. G. Stead, author of *Sharks and Rays of Australian Seas* (Angus & Robertson, London, 1964), but his questioning of the veteran fishermen involved left Stead convinced that at least a few gigantic *Carcharodon* still live in reaches of the abyss beyond the probes of man.

The person most impressed by the Montauk shark was Peter Gimbel, who had already seen two smaller whites that were caught from the *Cricket II*. In early 1965 he made a pilgrimage to the bar at Salivar's Dock, where the monstrous head resides today in a place of honor on the wall. He stared at the great shark and brooded over it and even dreamed about it, until it became a small kind of obsession. From this time forward, under the sea, he would peer fearfully about him, half expecting the massive shape to materialize in the blue mists. At the same time he longed for the confrontation, if only to exorcise a dread that anyone else would have thought extremely healthy. And in this wish, his film idea was born.

* * *

*Why does anyone in his right mind go looking for the biggest, most well-armed and aggressive cold-blooded animal in the sea? Because man is by nature curious . . . it is like every adventure in which man voluntarily pits himself against a challenging aspect of his natural world. Whether the struggle is to reach a mountain peak, to penetrate the most remote chamber of a cave, to explore the planets or space beyond, any explanation of motives is gratuitous. Nearly everyone has some of the curiosity of the explorer; hardly anyone would trade places with him, but nobody needs or wants his motivation explained. They feel it themselves . . .*

*This film tells the story of a search. We are after the most dangerous predator in the world, the Great White Shark. But . . . in order to find this creature in his own environment, we must cause him to hunt us. Where do we find him? Where does he find us? What will he do when we do confront him? If he attacks our cages and they yield, will our explosive weapons kill him with a single shot? We fear the thing we seek . . .*

The foregoing passage is from a treatment submitted to CBS by Gimbel for a film with the working title *Blue Water, White Death*: "white death" is an Australian name for the great white shark. Peter is embarrassed by the redolent prose that is required to excite the appetites of the entertainment world, but without it three long years of risk and research, including location trips to South Africa and Ceylon, would have been wasted for want of financing. As it was, his adventure seemed so speculative to the money people that the threat of withdrawal was held over him until the last frantic weeks before his film crew left New York.

FLYING BY WAY of Europe, the film expedition had arrived in Durban on March 30, 1969; I came down from Nairobi the following day. Most of the gear – ten tons of it, including the two shark cages and a recompression chamber – had been shipped ahead, and the remainder arrived on April 2. The *Terrier VIII*, a former whale-catcher chartered for the expedition, arrived from Cape Town on April 3, and the next two days were spent in arranging the gear aboard ship. A workshop was set up belowdecks in the former harpoon-rope lockers, and an electronics center for the film equipment in what was formerly the officers' mess. The last days before putting to sea were spent checking underwater camera housings for leaks, setting up deck lights for night filming at sea – the inside cabins had been painted titanium-white to reflect maximum light – assembling the submersion equipment and controls in the two shark cages, and countless other small preparations for a delicate operation under rough conditions in which one small mistake might cost a life.

The *Terrier*'s crew has been reduced to fourteen men, officers included, to make way for the film company, which moved aboard on the weekend. Gimbel and Stan Waterman, the producer and associate producer, are both divers and cameramen; the other two divers are Ron and Valerie Taylor of Australia, who have each won the Australian spearfishing championships three times. Ron Taylor, whom many consider the best diver in the world, is also alternate cameraman, and he and Valerie will operate the cages. Since people in the cages will be filming each other against the background of the undersea adventure, the four divers are the principals of the film; there are no actors. The director of surface photography is Jim Lipscomb, whose portable 38-mm camera will record the atmosphere of the expedition, the life aboard ship, the sea, the ports, the whaling operation, the ocean islands and, most important, the character of the divers between dives.

Lipscomb, Waterman, production manager Phil Clarkson, and Peter Lake, an apprentice film-maker who is responsible for still photography, are all veterans of Nassau. Lipscomb will be assisted by a young electronics engineer named Stuart Cody, who is responsible for the synchronous sound ("synch sound"); when Lipscomb shoots he automatically turns on the sound equipment, carried and monitored by Cody. This is the *cinéma vérité* technique, to be used here in place of conventional narration. Both will be helped by Tom Chapin, a strong young grip whose most crucial job is mounting Lipscomb's forty-pound Arriflex onto his shoulder harness and guiding him from place to place on the rolling ship, making certain that the precious equipment, not to speak of the director of surface photography, is not damaged in a fall. In addition to the harness, Lipscomb must carry a ten-pound belt of batteries, and he will operate whenever possible out of a chair on a platform with locking wheels.

Most of Lipscomb's film experience has been with television, but his film credits include *Storm Signal*, a fine documentary on heroin addiction that won first prize at the Venice Film Festival in 1967. He is sick of making movies for television – "One shot and then they're gone." He liked the white-shark-search idea and he liked Peter – "He's easy to work with, he's not dogmatic" – but most of all he liked the opportunity to make a full-length feature *cinéma vérité* film with a 38-mm camera. "It's never been done," he says. "*The Endless Summer* was shot with a wild camera and a narration dubbed in later. *Don't Look Back* and *Monterey Pop* were both shot 16-mm and blown up. This film offered a fantastic opportunity because there was enough money available to develop a 35-mm camera that could be carried on the shoulder." He showed me a special eyepiece with satisfaction. "Getting that thing over onto the side of the camera housing where I could look through it with the camera beside my head – that cost us four or five thousand dollars right there." Lipscomb was awed by the $750,000 advance that CBS's Cinema Center had awarded Gimbel for his first production, and a highly speculative production at that; he attributes it to Peter's no-nonsense manner, which people who deal in superlatives are not used to.

In the wings is Dennis Judd, sent out by CBS to make sure that Gimbel does not forget the commercial realities of his adventure. The danger is small: Peter is nothing if not a conscientious man, even when making preparations to risk his life. One evening in Durban he look me along to dinner with two of the local publicity people, who wished him to give speeches about white death, shake hands with seals, and generally play the fool on their behalf, all the while agreeing eagerly with his every protest. "Look," Gimbel said finally, "I'm here to make a movie and I'm up to my ass in work. I'll cooperate with you people for one press conference, but the publicity is *your* job."

The next day there was a press conference at quayside, complete with three models who moved as a trio, expressionless and inseparable as baby fish, and the papers next morning were full of silly stories about New York millionaires seeking white death in local waters. The stories curled the graying hair of Gimbel, who was portrayed in dockside photographs with Billie, his Norwich terrier and the film's mascot. Peter, who resents being considered a rich amateur simply because his father, Bernard Gimbel, was the head of the great New York department store, had three years of hard work riding on this enterprise, and he much disliked the aroma that this sort of cheap publicity was giving it. At the same time, he expected no better, and went on about his work.

On Saturday evening, April 5, Stan Waterman was given a modest party aboard ship to celebrate his forty-seventh birthday. Waterman is tanned and fit from years of working out of doors; he is a tall man, gray at the temples, with a monkish pate and a beatific smile of honest good will. Like Lipscomb, he is cheerful before all, and as this quality is also shared by Lake and Chapin, and intermittently by Clarkson, the small messroom rang with miscellaneous jollity. All these men are social animals, which is not to say that Gimbel, Cody and Taylor are antisocial but simply that they are less extroverted. Gimbel is chronically preoccupied; he lends his good nature and approval to group humor but rarely his whole presence. Cody speaks obliquely from behind thick glasses and a beard; he is the wittiest of the crew, but his humor has unpublic origins. Ron Taylor is even and pleasant, but he is a solitary man, and his quietness is so clearly in character that it never serves to dampen the rest of the company. As for Valerie, she speaks with spirit when she has something to say but rarely talks just for the sake of talking, and often sits quietly, holding Ron's arm; tonight she was quieter than usual. On the wall over our heads, beside a chart of the Indian Ocean floor, someone had posted a photograph of the great white shark underwater: the picture had been taken in South Australia by Ron Taylor, leaning below the surface from a platform attached to the transom of a tuna boat. The shark, which had come up in the wake, is gulping avidly at an oil slick, and beside the picture Lake had posted a small notice: YOU ARE WHAT YOU EAT.

The party concluded with a small cake, some champagne and a burst of song. Guided by Tom Chapin, a tall mustachioed young musician who plays the guitar and is an accomplished folk singer, Lipscomb and Waterman gave an enthusiastic rendition of "The Chivalrous Man-Eating Shark", which concludes:

Then they took her aboard in a jiffy
The shark stood at attention the while
Then he raised on his flipper
And ate up the skipper
And went on his way with a smile

*Chorus:*
The most chivalrous fish in the ocean,
To the ladies forbearing and mild,
Though his record be dark
The man-eating shark
Will eat neither woman nor child.

It was nearly a week since the crew had arrived in Durban, and everyone was restless. Because of the nagging unseasonal wind, the whale-catchers on which the first phase of the film depended had not sailed since Thursday night, when I had gone out on W-29. Also, the Easter weekend had closed the Durban stores from Friday through Monday, just when small last-minute pieces of equipment were needed, and the seldom-seen

Judd, without making fair allowance for these factors, had reported to New York that the film company was already four days behind shooting schedule. Within the company, small domestic disputes involving dogs and accommodation had cluttered the atmosphere, and Valerie Taylor, who was sitting beside me at the birthday party, made no effort to disguise her growing tension.

Valerie has the face of a sad angel, but she is high-strung and fiercely energetic: it is the energy that makes her a champion, for she is slight and feminine, with none of the coarseness that men like to associate with female athletes. Here in Durban she had spent the first three days feeling sick and the next three perishing of boredom: the men could work on the equipment, but for her there was nothing to do. "At home I'd be writing or gardening or painting," she told me quietly, her face tense, "but here . . . I haven't even got anything to write down in my diary!" Much of her resentment had become focused on the shark cage, and I remembered what Gimbel had said in the Bahamas about good divers being spooked by it. "I'm not going down in that thing," Valerie whispered, as if making a vow. "I don't care how many times they explain it to me. Why, the white death could snap that thing to bits, or take it straight to the bottom with him if he wanted!"

Valerie's diary, which she was kind enough to put at my disposal, would turn out to be extremely useful; it is candid, tart and generous all at once. Despite her illness of the first few days, the diary opens on an optimistic note. She had never been out of Australia before, and South Africa, she thought at first, was "really a great place. Very much like Australia in many ways though we of course don't have all these blacks wandering around . . . The blacks don't seem to be a problem, nor do they seem to have one, not with the whites, anyway." Also, she had a good impression of her fellow divers. Gimbel was "soft-spoken and seems kind", and Waterman was "even more gently spoken." (These qualities meant a lot to Valerie, who was put off by the loud voices at the mess table and the free use of four-letter words.) But she dreaded the killing of the whales: "The whole gory bloodthirsty business makes me sick – such is the human race in all its glory." Later, when she had actually witnessed the whaling operation, she wrote, "Sometimes I hate being a human." By this time she was also suffering her first doubts about South Africa.

While the ship waited for good weather, the shark cages sat on the foredeck of the *Terrier*, and on Sunday the burghers of Durban, drawn by news of adventures and white death, came down to the docks to have a look at them.

The Bahamas cage (which was ready in New York in case one of the new cages got smashed up) had been a great improvement on the Montauk cage, and the cages for the film were better still. The stabilizing mechanisms in the flotation chamber were more sensitive, and the aluminum bars were now inset in a heavier frame, so that the whole structure was stronger. To my unpracticed eye, however, the cages still looked sorely in need of a steel girdle which, spreading the impact of a large whitish body across the bars, would permit the cage to be pushed back through the water instead of splaying wide like a split fruit.

"Aluminium, eh?" A goateed Englishman named Basil Livingstone, who claims to have caught white sharks at Saldanha Bay, inspected the delicate bars and shook his head. "Seems a bit flimsy to me, I must say." I assured Mr Livingstone and myself that the delicacy was required for the suspense and aesthetics of the film; a great clanking jail cell, impervious to attack, would not be as exciting to an audience as this fragile device. Besides, it would have to be connected to the ship by a frightful great steel cable – eh, Mr Livingstone? – whereas this lovely thing was free as a bird. Mr Livingstone looked doubtful.

We stood about shaking the bars for a little while. Livingstone is the owner of a whole fleet of small vessels, mostly crawfish boats (African rock-lobster tails), and his fleet includes the *Terrier VIII*, which is used primarily for research, salvage and general charter. She was built in 1951 in Norway and, as a whale ship, sailed out of Cape Town as recently as October 1967; before that she was thirteen years in the Antarctic. Most of her crew are whaling men, but the passing of the whales makes it unlikely that she will ever hunt again. At 158 feet, she is slightly smaller than the Union Whaling Company vessels, but in general appearance she is much the same: high in the bow and very low amidships, with an outsized smokestack for her 2200 HP steam engines. Her cruising speed is 12 knots, and her range is four thousand miles.

As on the whale-catchers, the officers are Norwegian, and the crewmen mostly half-castes of the category known here as Cape coloreds; these men have the South African cockney accent, and some could pass for limey sailors in any port in the world. The whalemen among them are fascinated by the search for the great white shark. Like the crew of W-29, they claim that they have never seen it in these latitudes, though it occurs farther south in the cold waters off the Cape of Good Hope. The *dangerous* shark, they say, is the Zambezi shark, which swims into the Zambezi River, to the northward; though its classification is disputed, most authorities believe that the Zambezi shark is *Carcharhinus leucas*, which enters fresh water in the Tigris and Euphrates, in the Ganges, and elsewhere; it occurs as well in Lake Nicaragua and in other lakes in Guatemala and New Guinea. In all of these places it is dangerous. Yet the three biggest specimens of *C. leucas*, which is known on the Atlantic coast of North America as the bull shark, have been taken in Chesapeake Bay, and it turns up also in Lake Okeechobee; in neither place has a shark attack ever been recorded.

The opinion of the crewmen must be taken seriously, since they have seen thousands of sharks, though whether they have looked at them is another matter: when closely questioned, they admit that they have never seen the great white shark at all, in cold waters or warm. "To us, a shark is a shark," says Jan Moen, the acting captain of the *Terrier*. From a boat, most people would have difficulty distinguishing a ten-foot white from any one of several species, since the color of sharks – the blue shark is an exception – varies greatly within its own kind, and is generally undistinctive. Also, the mass of any fish is difficult to estimate from above the surface, so that a ten-foot white and a ten-foot white-tip would be chiefly distinguishable by their fins. Due to its great girth, a large white shark, twelve feet or more, is unmistakable, but since it is a rare fish, and since the common tiger shark may exceed this length (though not the girth), a shape seen indistinctly in the murk around a whale might be ascribed to the latter species.

* * *

By Sunday afternoon the wind had moderated, and at 3.30 on Monday morning the whale-catchers left port; the *Terrier VIII* put to sea an hour later. At daylight she was headed eastward into a big gray ocean day, with big gray swells: the wind was northeasterly at 15 knots, and the seas loomed one by one beneath the hull and rolled away to the southwestward. From the bow as it lifted and fell, parting the waves, came a loud wash of water, and albatrosses, wandered north from Antarctic latitudes, curved back and forth across the sky.

Breakfast was scheduled for 6.30 a.m., but not everyone felt up to it. Among those who appeared, most ate lightly and two had to take their food on deck in order to eat at all. Despite this, the spirits aboard ship were high; even Peter Lake, who felt worst of all, was able to be humorous about it, and by evening he was much better. At twenty-five, Lake is heavyset in the way of someone who has not yet lost his baby fat, and at this stage of his life wears round glasses, three cameras on his chest, mod clothes, hair and a mandarin's mustache; a jolly fellow, bright and generous, he is unkempt in every way.

The *Terrier* followed W-29, which is the lead boat of the fleet and has the best electronic gear. Jan Moen kept radio contact with Willy Christensen on W-29, but it was late in a long morning before the first whales were sighted. I spent most of the morning on the bridge looking for spouts, and Stan Waterman joined me; he talked for a while about the coast near Camden, Maine, where he once had a blueberry farm and still owns a summer house.

Like Gimbel, Waterman had never seen a white shark underwater, but last summer one came up and nudged his skiff in the Gulf of Maine: since the shark was longer than his boat, he was considerably impressed. Not long before in this same area, two boys had barely been rescued when a white shark attacked their boat, and another shark had demolished a dinghy that was being towed. In 1953, off Cape Breton Island, a lobsterman drowned when a white shark foundered his dory; his companion clung to the wreckage and survived. White-shark attacks upon small boats have occurred regularly in the Maine–Nova Scotia region, and apparently the fish is noted for this tendency wherever it is found.

I asked Waterman how he could tell that the shark was a great white, and he said, "I don't know – I just *know*. If you are familiar with other sharks, and then this one comes along – maybe it's just a process of elimination, but you know you're seeing something different." Ron Taylor is more explicit. In his opinion, the conical snout which gives the shark its Australian nickname "white pointer" is a diagnostic character even from a boat (the tiger shark has a broad rounded head), and he says that sometimes the white's big teeth, protruding like those of a crocodile, are actually visible from above the surface. Underwater, even a small white shark may be instantly distinguished by its great black eyes, like holes; no pupil is visible. There is also the crescent-moon mackerel tail, for which the swift mackerel sharks, Isuridae, are named, and a peculiar down-curve at the corners of the mouth which gives this beast, when observed head-on, an odd sort of wistful smile.

Valerie Taylor agrees with Waterman that something intangible makes this species unmistakable. "You can *always* tell him," she told me once, "even from the surface. It's his whole attitude. The first time we ever saw one, we knew it was something different, and it was just swimming toward the boat. It *moves* differently, like a torpedo; there's not all this thrashing back and forth like the others do. You can tell it a long way off, the way it comes gliding in, and the rest just get out of the way. Other sharks pass back and forth and never look at you, but not him. He comes up beside the boat and turns on his side, and that big black eye comes right out of the water and looks at you. Oh, he's *intelligent*, he is – *that's* why you have to be afraid! Sometimes he stands on his tail in the water, stands there for *seconds* sometimes, clean out of the water with his gills showing, to get a better look!" She was rubbing her arm. "You see? I'm getting goose flesh, just talking about him!"

The white shark's reputation for attacking small boats in order to get at the occupants suggests more intelligence than sharks are normally granted, and it seems most unlikely that it elevates its head to reconnoiter. While such vision is ascribed to the orca, or killer whale, which is highly intelligent, even for a mammal, the shark eye is an underwater mechanism that in the air may see no better than the naked eye of man beneath the surface. Still, this species appears to have a taste for warm-blooded mammals – cetaceans, seals and man. Possibly Valerie

Taylor's simple explanation is as good as any: "He's big, and he needs big food."

Toward noon, the *Terrier* came up on a harpooned whale marked with a yellow-red-yellow flag. Lavender entrails, extruded from the harpoon wound, washed back and forth, and a gang of albatross, rising and falling on the swells that washed the carcass, plucked at the guts, haggling over the freed bits with the hollow, horrid gobbling of seafowl.

At the prospect of action, Valerie Taylor's spirits were restored; she was anxious to sit on the back of the whale to give Peter Lake some publicity shots. Gimbel refused permission, not wishing to take any needless risk with equipment and personnel. He moved methodically about the deck, checking equipment, and although no cameras were to be tested, it was three in the afternoon before tanks, wet-suits, weights, regulators, fins and leg knives were dealt out, and the trial cage was swung over the side.

With her fifteen-foot draft, the ship was steady in a stiff afternoon breeze, the cargo boom worked smoothly, and the swinging cage with Gimbel in it, held off the hull with boat hooks, was lowered without incident into the sea. The cage was tethered to the whale's toothed lower mandible by the waiting dinghy, and then the divers, one by one, were given their first lesson at the controls. Ron Taylor was first, then Valerie, who entered the cage without hesitation and operated it very well. Stan Waterman was scheduled next, but he was busy in the dinghy, and I took his turn; when Clarkson took over in the dinghy, Waterman replaced me in the cage.

Though no sharks were visible from the surface, we all saw sharks beneath, which made me wonder how many sharks I had not seen on my voyage on W-29. However, I made the worst of my opportunity. I was still a novice, and this was my first dive since the Blue Hole, eight months before, and besides trying to manage a leaking mask and a broken fin strap – an escaped fin would drift down for over a mile – and ear trouble and general apprehension, I was also attempting to operate the cage. As it turned out, I managed the cage perfectly well, but I could not get a hand free long enough to tip my mask and rinse it clear; a misted mask seemed the least of my multiple problems. At one point, however, seeing some small pilotfish swim through the cage, I pointed out this sign of shark to Peter, who turned immediately in the direction they had come from. What he saw was a twelve-foot oceanic white-tip shark, which he later estimated as over a thousand pounds. It went cruising past not thirty feet away, he tapped my shoulder and pointed, but all I saw was a dim shadow. I missed completely a small tiger shark that came and went a few moments later. By the time I got organized enough to rinse my mask there were only a few two-inch sergeant majors (though I found this extraordinary enough: what were small reef fish doing in the open ocean, seventy miles from the nearest shelter?) in a beautiful pale-blue light full of shimmering small jellyfish, some of which stung Phil Clarkson painfully as he handled the dinghy lines on the surface.

Waterman later reported a strange jellyfish that wriggled along horizontally like a fish; he made a sign to Gimbel that he wished to leave the cage to have a look at it. Having seen big sharks, Gimbel was doubtful, and a moment later a big blue shark came in out of the mist. Blue sharks are bold, and this one came right up to the cage. Gimbel tapped Waterman, who was looking the other way. Confronted suddenly by a big shark at close quarters, Stan gave such a violent start that he actually struck his head on the cage roof, eliciting a good deal of soundless mirth. Later, while the cage was being towed back to the ship, Gimbel saw four more sharks, and back on board he was exultant. Not only had the ship and cage worked well together in marginal conditions but there had been sharks, big sharks, of three species, around a whale that had hardly bled at all. Valerie had relented about the cage – "Oh, I didn't like the look of it at all when it sat on the dock, but it goes down nicely, doesn't it!" – and the first day was considered a great success.

I was having a drink with Peter in his cabin overlooking the foredeck, when Ron came in and joined us. "I *know* the great white's here!" Peter told Ron. "I can just *feel* him!" Taylor stood silent for a moment, considering this. He is a well-made graceful man with long sideburns and a black monk's cap of unparted hair, a flat gaze – his eyes do not open the way into his mind, but reflect one's own – and a slightly retracted lower jaw; perhaps it is association, but in a strange way that eludes definition, Ron brings to mind a shy and handsome shark.

Finally he spoke, in his measured Australian cockney.

"Yis," Taylor said. "'E should be."

At supper the talk was about sharks, specifically the great white shark, which is still an almost legendary beast. Once notorious as the great swift shark that followed sailing ships for days on end, it was "to sailors the most formidable of all the inhabitants of the sea," wrote Jonathan Couch in *Fishes of the British Islands*, "for in none besides are the powers of inflicting injury so equally combined with the eagerness to accomplish it." This opinion, never disputed, sums up most of what is known about the species.

The white is the only shark that Ron Taylor is afraid of. "I'd swim with one," he says, "but never two." Sharks cannot stop short like fish, much less back up, and Ron feels that one shark can be maneuvered – "You can push him off" – and, if necessary, shot: he carries an underwater gun of his own design that fires a large .303 slug. Flight is useless; it is thought that sharks of the family Isuridae may swim as fast as forty miles an hour, though probably they rarely do so unless their prey impels them to that speed. Frank Mundus, the shark fisherman of Montauk, once told Gimbel that in spite of its bulk the white was the swiftest and most agile shark he had ever dealt with, not excepting its relative the mako, which is the only shark in the world classed as a game fish.

*C. carcharias* is an open-ocean creature in most parts of the world, and the few taken alive have died of shock or self-inflicted damage in their tanks before much could be learned about their habits. White sharks that have strayed inshore may be rogue sharks, too old or disabled to compete in the open ocean; in any case, they must be regarded as extremely dangerous. This species is thought responsible for both the celebrated series of shark fatalities – in New Jersey, in 1916, and in California in recent years – that have occurred in this century in the United States; in Australia, the creature's common names, "white death shark" or "white death", are due to the number of fatal assaults on human beings. The coast of southern Australia is one of the few places in the world where the species occurs commonly inshore, and it is there that three friends of the Taylors were bitten. (Shark attacks are rare in water of less than 70-degree temperature, but white sharks seem no more inhibited by cold water than by anything else: the California attacks occurred in 55-degree water, and in South Australia the water was in the 60s.) In 1967 a fourth friend, Bob Bartle, was bitten clean in half by a white shark, off the west coast of Australia near Perth. The shark swam off with his lower half, and Bartle's head and trunk were retrieved by his partner, who dragged the remains out onto the beach and photographed them before going to report the death. This foresight was wasted, however, since the pictures were so grim that no one would buy them. "That was a chap I never liked," Valerie says, "even before that happened. He's a good diver, though."

In South Australia some divers prefer bright colors to black wet-suits, which tend to give man the appearance of the sea lion that is an item of the white shark's diet. But in airplane crashes at sea during World War II, as Ron points out, the gunners in orange flight suits were attacked far more often than the pilots in dark blue. (Orange, the color required by the U.S. Coast Guard for life jackets, is visible for great distances at sea, but since a shark's vision is limited in terms of distance, though it has very acute light sensitivity, there seems to be no immediate connection.) Here again, very little is known; when it comes to self-protection, each diver must go on his own experience and instinct.

THE EVENING WEATHER report forecasts marginal weather. Since most of the days have been much worse than marginal, we decide to lie to at sea and try to work again tomorrow.

April 8 is a clear blue day. Great swells of vanished storms persist, but the breeze is light. Overnight the *Terrier* has drifted shoreward, and daylight finds her forty-five miles off the coast.

By 7.00, two whale-catchers are in sight down to the south and the *Terrier VIII* is headed southeast to cross their bows. Then Willy Christensen comes on the air: W-29, forty miles to the northeast, has sighted whales. Our ship alters course, rolling northeastward at 12 knots. Off the starboard beam, a solitary albatross bends the sky in hard black-and-white arcs. In the morning light the mist from the bow wave makes small fleeting rainbows: the boat's wash is pale turquoise where the white foam pours across the deep stone blue.

At midmorning, W-29 and W-18 rise on the skyline; the spotter plane crosses the *Terrier*'s mast. The flat voice says that W-29 has whales on her sonar; she is barely making headway, ploughing a soft furrow in the sea. Ahead of her the ocean parts, and eight whales, glinting, blow together and subside before their spume has settled on the waves. But the winded animals soon surface, and now W-29 has picked up speed, her big stack shooting filth into the sky. She rides down upon the whales so fast that after the shot the one struck falls immediately astern; the ship throws her engine into reverse, and the boiling wake turns red. The whale, wallowing, is hauled in by the big winch and given a killer harpoon; the harpoon with the air-hose point pulls out before the crew can noose the churning fluke. A third shot resounds, then a fourth – the explosion smoke is pink – and finally the whale is still.

Quickly the blood spreads until an acre is reddened, then much more, and just as quickly the sharks come, attracted as much by the repeated explosions and the thrash of the dying whale as by the blood; low-frequency vibrations picked up by the nerve endings in its lateral-line system may bring a shark from hundreds of yards away, and even a very small amount of blood, spreading out through millions of gallons of water, may be scented a quarter of a mile down current.

The first shark appears before W-29 has released her marker buoy, and in minutes there are six in sight, then nine. Before the *Terrier* comes alongside and hauls the yellow flag aboard, there are twenty or more tawny white-tips gliding in. At first their attack is tentative; a vulnerable wound has not been found. Staring down from the bridge, Jan Moen remarks, "That sperm hide is tough; if that was a baleen whale, now, you'd see big chunks out." Moen, a decent man with a sudden boyish smile, once served as mate for Torgbjorn Haakestad. Because he doesn't have full master's papers for ships operating offshore, the *Terrier* also carries a licensed master named Knut Paulsen, who is rarely seen; perhaps he has confined himself to quarters. Officially Moen is the mate, though it is he who operates the vessel.

Sharks bite and churn, rubber bodies twisting, and fall away again; the fins flop on the surface. Then a big hole is opened up behind the head and a brief frenzy takes place, eight or more sharks rolling in the wound; this awful motion, in which the shark fills its jaws, then rolls completely over on its back to twist the gobbet free, is known as "coring", as in coring an apple. But the rest do not join the frenzy, which soon subsides; there is plenty for all. The sharks attack in twos and threes, hour after hour, feeding steadily at the widening trough.

Birds come too: the wandering albatrosses and a smaller albatross, with black bill, dark mantle and dark edges fore and aft on its underwing – probably the yellow-nosed albatross (*Diomedea chlororhynchos*), named for the tip of its beak – and sooty shearwaters and storm petrels. A jaeger goes hawking past, too wild and shy to settle on the whale. In the ocean emptiness, far off to the south, a great fish leaps high against the sky, standing on end, and falls back into the sea – a marlin, or a mako . . .

* * *

On the foredeck, the men rushed to get the equipment overboard. The haste was precipitated by the horde of sharks, which filled the whole ship with nervous energy, and the most energetic of all was Phil Clarkson, who was responsible for getting the operation under way. Clarkson is a small, well-knit man, very handy and quick and impetuous, with a sharp hectoring voice ("I hate my voice," he announced once at breakfast – the kind of bare candor for which one forgives anybody anything), and his difficult job was made more difficult because he has not learned how to give orders to his peers. ("Phil still thinks he's talking to those poor migrants on his tree farm," one man grumbled.) Though the two are long-time friends, Clarkson clashes frequently with Gimbel, who is painstaking and methodical and easily distracted by details; Clarkson feels that if he does not keep the heat on Peter, the operation would never get under way at all. Today, however, Phil was being efficient to the point of inefficiency, and in hurrying everybody else, went off half-cocked. Efficiently, both cages were launched without having their systems checked – all the more serious because both were new, and one had never been in the water. The other did not even have its air cylinders hooked up, and had to be efficiently hoisted out again.

When this was discovered, Clarkson was out in the rubber dinghy, and Gimbel, wrench in hand, stalked over to the rails. Peter is tall and balding with a big open generous face and a big chest and shoulders scarred by operations, and when he is angry, his voice, which is soft and husky, resounds like a foghorn. "Look!" he yelled. "I can't make it with this hustling! You hustle in this game, Phil, and somebody is going to get killed! What's your hurry, anyway? It's not even noon yet, we have good light and those sharks aren't going any place – if they do, they'll be replaced by two to one!"

Clarkson took this very well, and after that, considering all the things that could go wrong, preparations proceeded smoothly. The main impediment was sharks, which were milling in such numbers that in the rough surge the transfer of divers from rubber boat to cage was a nervy business. The sharks nipped and brushed the boat repeatedly, and Phil later reported that the outboard had been stalled four times by sharks biting the propeller.

Rather than take any more time, Gimbel decided to check out the unchecked cage in action: this was not foolhardy, since the cage would remain tethered to the whale by a twenty-foot line. Ron Taylor ran the cage on the first descent, and Peter took one of the wide-lens Arriflex cameras fitted to a Mako submersible housing: the action was so fast that a ten-minute magazine was burned up almost as fast as he could shoot. Since the South African government had not yet issued ammunition permits for the "bang stick" – short clubs like police billies, tipped with powerheads that discharge a twenty-gauge shell into a shark when jammed against it – Ron carried his sling gun that fires a .303 bullet. "I reckon this might even stop one of them *whyles*," he said, "if I put it in the proper place."

Ron came back aboard the ship while the camera was being reloaded, and I went down with Gimbel for a look. In the transfer from dinghy to cage – we were entering through a small emergency hatch in the roof of the cage, which at the surface is awash – the boat slid down the back side of a wave, and I found myself stretched between cage and boat; I chose the cage, attaining it in a single violent motion. To judge from the cheering aboard ship, no one had thought that a man in a heavy scuba tank could be so agile.

Shark tails flopped like lily pads on every side as the cage sloshed up and down, jarred by the black gleaming carcass of the whale. As Peter bubbled impatiently below, I sat in the warm water for a moment, adjusting my gear, then opened the hatch and slid into the limpid 79-degree water. Gimbel, who hates the surface motion of the cage in a rough sea, took it down immediately to twenty feet, and the contrast between the glare and slosh and bang and clamor at the surface and the blue stillness beneath was as sudden as the thermocline at the Blue Hole.

Looming overhead, the black oval of the sperm whale broke the mercurial silver sheen of the ocean surface. On the surge of passing seas, the tail fluke waved in the wet sun, graceful in death, and the long jaw that had once seized giant squid deep in the abyss hung slack, teeth silhouetted. By hauling on the line tied to the jaw, the cage could be drawn beneath the wound back of the chin. Into this great cave of flesh the sharks bored in twos and threes, tails lashing in the bursts of gore that flowered in the sea; the shredded flesh, a bright dead-white, rippled like soft corals in a current. The cage itself was shrouded in red murk in which shadows thickened and dispersed; then the water cleared again, and a jellyfish glistened in the sun shafts a hundred feet away.

The cage had entered a blue realm of dream. The water glistened as if filled with its own light, and the utter silence made the scene more awesome, a nether world of open-mouthed dead staring forms that moved in slow predestined circles: having no air bladder to buoy them and lacking a true breathing apparatus, most sharks must draw oxygen from the water pouring through their opened jaws and washing over their gill surfaces, and are doomed to keep swimming, open-mouthed, from birth to death. Because they do not weigh much more than the water they displace, their movements seem effortless; they glide forever through the seas like missiles lost in space.

In the blue void the big shapes were so numerous that one could not turn fast enough to count them; they moved in concentric circles and in opposite directions, smooth as parts of a machine. There were certainly more than forty sharks – the divers estimated fifty – none of them smaller than ourselves and many ten feet or better: blues, duskies, white-tips and a mustard-colored species with a strange curled dorsal fin that looked from a distance as if the whole top part of it had been sheared off. Just beneath the flat edge was a distinct lateral ridge that ran the whole length of the fin. None of us had ever seen or heard of such a shark.

Some of the white-tips and duskies were twelve-footers, heavy, businesslike brutes with glazed yellow eyes, a tawny hide and long back-swept pectorals tipped with luminescent white; the blues are thin, with pointed snouts and small black eyes in which the pupils are quite visible. These three species and the tiger shark seen the day before are requiem sharks, *Carcharinidae*, and all four are included in the list of nine established man-eaters (among 250 known species of sharks) that is one of the few things agreed upon by shark authorities. One ten-foot blue nosed repeatedly at the cage; when I kicked it in the side, it slid away. Valerie stroked a passing white-tip and Ron punched one in the throat; like the blue, they fled. Another thing shark authorities agree on is that sharks should never be provoked, but probably the exact reverse is true. As Gimbel says, a shark has no mammalian responses. "This idea that you mustn't provoke them is nonsense. If you really hurt one, he'll go find something easier to eat." (This was also the belief of Pliny the Elder, but the theory has fallen into disrepute in the twenty centuries since.)

Some of the sharks carried brown remoras, which attach themselves to larger creatures by means of a disc on the top of the head, and all were accompanied by pilotfish, a small member of the jack family with vivid striping, blue on black. The visibility was well over one hundred feet, and the pilotfish shimmered like striped petals as far away as one could see the sharks, which formed and vanished in the ocean mist. From below, the long-winged white-tips were surreal, silhouetted on the silvered sun like ancient flying beasts.

When water and body temperatures are so nearly the same, the skin seems to dissolve; I drifted in solution with the sea. In the sensory deprivation of the underwater world – no taste, no smell, no sound – the wild scene had the ring of hallucination. The spectral creatures came and went, cruising toward the cage and scraping past with lightless eyes. Then I remembered the great shark we were awaiting, and peered past the coasting forms to the region, three whales' lengths away, where the blue gathered into mist, or down where the blue turned to night, awaiting the colorless heavy shape with eyes like black holes in a shroud and a slack smiling mouth. I too sought as well as dreaded a confrontation with this dispassionate unblinking death which is part of life.

"I have never seen so many sharks in my life," Valerie wrote later in her diary. "All big ones, too . . . without the cages, we would not have lasted 10 minutes." At one point Valerie tugged at Peter, who was leaning out of his cage to photograph: a big white-tip had gone into a glide straight at him, veering off only when Gimbel whirled. "I have a healthy respect for the old white-tip," Gimbel commented. "He's a good strong husky shark, and he's the first shark that ever attacked me." Like most pelagic sharks, these requiem species are widespread, and both blues and white-tips are common off Montauk, where Peter's first experiments with the cage were carried out.

Already Peter was wondering if the divers should try to leave the cages. "Off Montauk," he told Jim Lipscomb, "those blues would bump into you while you were filming and never bite, but I became convinced that they would bite a moment after the bump if they didn't get a strong reaction to show them that there was still life in what they had their eye on. I think that these open-ocean sharks here in the Indian Ocean will behave the same way. I'm not a big enough fool to jump right out of the cage with the certainty that I'm right, but we'll watch them and see how they behave. Eventually we'll get a feel for them, and if they behave in the same pattern as the blues, we'll work them the same way."

By 4.30 it was too dark for underwater filming, and W-29 was already on the horizon, coming to fetch her whale. W-29 taking the whale in tow would make a fine end to the day's sequence and I went to alert Lipscomb, but Jim, seasick, had been lugging his heavy rig since dawn and he was exhausted; he sat wall-eyed on his bunk, unable to move. A little later, however, he made it out onto the bridge, shooting the whale-catcher in a marvelous dramatic twilight as she hauled in the big bull with the yawning wound where the left flank had been and steamed away westward toward the coast.

On deck, we lashed the cages down, hosed the salt water off the diving gear and stored the cameras. None of the divers had ever seen such an aggregation of sharks, and the talk was excited; in the old-fashioned speech that he favors when in fine spirits, Stan Waterman called out to me how lucky I was – "You *are* in luck, sir!" – and I agreed. I was delighted by the experience and delighted it was over, and when Valerie brought whiskies around, I took mine up onto the gun platform, already drunk with exhilaration. The death of the day was beautiful, and as the *Terrier* came up with W-29, my spirits soared with the great white birds crossing her wake, lifting and falling, tipped on end, dipping their stiff black wing tips into the sea.

Ashore, the publicity people made the most of the shark packs that had surrounded our "intrepid divers" on April 8, and one paper had a story about how Ron Taylor became "terrified" when a ten-foot shark became entangled in the line that tethered the cage to the whale: "I thought the cage would break up," he was quoted as saying. The story annoyed Valerie very much – "I've never known Ron to be terrified in my life!" she exclaimed at supper – but her husband was unruffled. "I'll sue them for character assassination, I will," he murmured, to nobody in particular.

"Ron's always under control," Valerie says. "He's never lost his temper with me, and I've never seen him show excitement. It's his temperament and his philosophy: whatever you do, you'll get it back, you reap what you sow; if you hurt someone, you're going to be hurt, and the same thing about honesty and everything. Oh, Ron's a beautiful person, and he's intelligent. He can turn his hand to anything, except my garden; if I ask him to prune a bush, he looks like he's been struck by lightning.

"Anyway, his calm is what makes him a great skin-diver. He's got a big lung capacity and he knows his fish too, but it's that calm that makes him a champion. But winning never meant much to Ron after the first time.

"Now, I'm quite different. As a diver, I have to get by on nerve, because I'm not even a good swimmer. I have a good lung capacity now, but I had to work on it. I don't win competitions because I'm a good diver; I win because I'm still going at it like a mad dog when the rest have been hauled out panting. I'm not even calm; I weep when I get frightened." She described how once, during a night-filming trip on the Great Barrier Reef, she went in first and had descended the face of the reef when her light went out. "I just clung weeping to the wall. Ron never came and never came, and I thought he had abandoned me. Finally I got up courage and crept up the wall – it was pitch-black, I kept bumping into unknown shapes – and finally I got to the side of the boat. And Ron says, 'Here, just hold on a minute, Valerie, and I'll fix that light for you.'"

Ron Taylor and Valerie Heighes met at St George's Spearfishing Club in Sydney when she was twenty-one. Ron's interest in diving had begun at the age of fourteen, when he found a face mask in a Sydney tide pool; Valerie had been brought up along the coast, and had been married briefly to a skin-diver. She was a commercial artist and an actress, and had once starred in a production of *The Seven Year Itch* that ran for nine months. "I was never a great actress, but I had heart. I'd drop everything and go anywhere to get a part. Also, there's a shortage of scriptwriters in Australia, so I started writing TV shows that had great parts for myself, like being thrown into a shark tank at night by the villain. Then I'd go in and apply for the part of Paula Howard, and they'd say, 'You'll *have* to play it – nobody else would touch it.'

"But Ron paid no attention to me at all. He was a photo-engraver then, very shy and quiet – he'd never had a girl friend. I tried everything. I'd challenge him to ping-pong and then turn out in a bikini, jumping around to get his eye. And when we were diving with the club, I'd always sneak around ahead of him; every time he came around a rock, why, there'd I be. I was *always* right where he could see me, and I wasn't the only one, either."

In 1961, when Valerie was twenty-five, they started working together, and they were married three years later. "I finally said, 'Either marry me or I'll marry that folk singer in two weeks!' And to everyone's surprise, he married me."

After their wedding, Ron and Valerie went into diving and underwater photography on a full-time basis. "I was always willing to try anything, I used to drop jobs right and left, but for Ron, making the break was harder – he'd always been very careful, and he wasn't used to change. And it *was* hard at first – we had to peddle the fish we speared to pay for our film. But now things are better. Ron's made some beaut films and we have our own boat and we give talks – oh, and that's where we're different too: I'm always champing to get out there and get at the audience, but he gets very nervous. He thinks I'm a show-off. Like when I shot sharks for *Vogue* wearing false eyelashes." She shook her head. "No good asking Ron what makes me a good diver – he'd just laugh. He thinks I'm *not* a good diver because I'm too emotional, I just rush in where angels fear to tread."

Valerie left, but in a little while she turned up again with a magazine article about herself. The article, which described her adventures with sharks, portrayed Valerie standing beside her own paintings of a clown, flowers, fish. "That's what I *really* am," she said, "a painter."

Valerie's near-perfect face is almost invariably intense, with an anxious shadow at the eyes; she touches Ron every little while for reassurance. The Taylors are childless and, more than most people, they pronounce each other's names – Ron? Yes, Valerie? Don't you think so, Ron? – like children keeping track of each other in the dark.

* * *

One reason for the *Terrier VIII*'s return to port was the need for the portable powerheads. Gimbel wants his divers to carry a bang stick on each descent, because if a big shark attacked the cage, or a cameraman was caught outside it, this protection would be crucial. But two days later, when the *Terrier* put to sea again, the permit for powerhead ammunition had not come through. South Africa is very nervous about permits for firearms, even underwater ones. "Rumors of unrest are exaggerated," an article in one paper stated, "but in any case, whites control all the centers of power: revolution is inconceivable . . ."

Except to swim in the surf and walk the beaches, I rarely left the hotel; I even gave up the idea of visiting the famous Kruger Park. I was content at sea, beyond the territorial limits, and depressed upon each return to shore, and was therefore especially sorry when, on April 9, the northeast wind, "the bluebottle wind", set in again, keeping the *Terrier* in port; it brought white swatches to the dirty sea and a plague of small blue jellyfish onto the shore. Like everything else in Durban, the beach is segregated into "white", "Asian and colored", and "Bantu" segments of very different size and desirability, but below the tide line the people cross the barriers, hurrying across the desolate gray flats. On the day of the bluebottle wind, the tide line was littered with the elevated air bladders, like tiny blue balloons, and dark figures, crying out to one another, danced from jellyfish to jellyfish, popping the balloons with hard small angry stamps of their hard heels.

On the strength of the successful shake-down cruise, the Durban papers published a new article on Gimbel, in which it was said, among other things, that his abandonment of Wall Street for a life of adventure had begun with the death of his twin brother, David, in 1957; the New York millionaire realized that life was short and decided to live it "to the hilt" from that point forward. The truth is that before David died Peter had already made his adventurous dive on the *Andrea Doria*, and after the death he had remained an investment banker for three more years. "I was on Wall Street for nine years," he told me, "and while I was doing it, I really thought I liked it." But the challenges of Wall Street are well charted, and he did not want to go into business with his father. "I wanted to be a success," he says, "on my own terms."

In his Wall Street years, I did not know Peter well, though I saw him occasionally in New York. I lived far out on eastern Long Island, writing in the winters, working as a commercial fisherman in fall and spring, and running a charter fishing boat out of Montauk in the summer. On a July day in 1956, Peter had chartered my boat; he and his party were to be at the dock at 6 a.m. By 6.30 his friends had come, but at 7.30, Peter was still missing. Later we learned that, the night before, he had flown up to Nantucket, where the *Andrea Doria* had just gone to the bottom; he had hired a boat and made the dive that made his reputation as a diver.

Another year, in the spring of 1959, I think, we both went parachuting in Massachusetts; this was the first wave of the sport of skydiving that has since spread across the country. Alone among my friends who tried it, Peter applied himself seriously to parachuting and later put it to good use; in 1963, he and three others (one of them was Peter Lake) made a risky jump into the wild highlands of the Peruvian sierra and worked their way down to the Rio Picha, in the headwaters of the Amazon, in an unsuccessful attempt (later described in the *National Geographic*) to find some ancient ruins that he had read of in *The Cloud Forest*, a journal of South American exploration written a few years earlier by myself.

At the time of his Peruvian journey, Peter had already left Wall Street. To the great disappointment of his father, who was proud of his son's adventures but refused to take them seriously as a way of life, Peter enrolled at Columbia University, intending to study for a degree as a biologist. In 1964, with Carleton Ray, he dove beneath the Antarctic ice to film the Weddell seal. He was still caught up in the mystique of sharks and, in 1966, invited me to a screening of his excellent short film on blue sharks. By this time he had lost interest in the biology degree, his marriage was in disrepair, and his life, which seemed so colorful to others, had begun to strike him as directionless. He was desperate to find a living which combined his interests – science, diving, film-making, adventure – and what evolved not long thereafter was the project that became the search for the great white shark.

ON THURSDAY, APRIL 10, the weather was still marginal, but before dawn the next morning the whale-catchers put to sea. The wind, moderate at daylight, had risen by noon to 30 knots. The mist of whale spouts could not be seen in the spray that blew from the crest of every wave, and it was just as well that none were found, because the light was poor and the film crew was seasick, and launching the cages would have been dangerous. In the late afternoon the *Terrier* itself encountered a pod of whales, and W-29 came churning over, throwing spray like great white wings, but the whales were all small cows with young, and none were taken.

The next day the wind had died to a light breeze. Word came early that W-I7, far down to the southward, had harpooned seven whales as fast as she could handle them, and Gimbel, on the bridge, exclaimed to Jan Moen that W-17's Captain Arvid Nordengen, whom he had accompanied on a whale voyage during his location trip the year before, "never missed". Moen nodded politely, but after a few moments he observed that good gunners missed quite often. "They take chances that other gunners do not take," he said, "and so they miss, but still they get more whales." Since Nordengen planned to gather his whales immediately and tow them into Durban, the *Terrier* remained with W-29. Captain Haakestad's ship, with her big number on the stack, had already been filmed; for purposes of continuity and flexibility when the time came for cutting and splicing, it seemed best to stay with her.

At 11.30, W-29 located three big whales, and the *Terrier* caught up with her at 12.05, just as the three bulls broke the surface ninety miles offshore. Within the hour, all three were floating beneath flags. The first was killed by Haakestad's first shot. It scarcely thrashed, and there was no blood at all. It was a big bull, and as it lifted and sank in the great swells of yesterday's storm, an enormous appendage, extruded *in extremis*, lashed about violently like a live thing trying to bore into the belly. Glimpsing this as she climbed onto the gun platform, Valerie cried, "What's *that*?" before she thought. I gazed at it sternly. "That's his penis," I said. "*Oh*," she said, "I thought . . ." Chagrined, she did not bother to finish. Mere words were superfluous; her raised eyebrows said it all.

Soon two sharks appeared, but they kept their distance and were not joined by others. Off to the eastward, the second bull had already been killed; it too had attracted two sharks, but it had been shot twice and was bleeding, and there was hope for more.

The *Terrier* was tied to the second whale by 1.15, but two hours passed before the divers went over the side; one of the underwater cameras didn't sound right, though it seemed to be functioning, and there were other small delays. I had a certain sympathy for Phil Clarkson, who stoically held his tongue. There was a big difference between being hustled and being ready to enter the water when the chance came, and for selfish reasons – the longer we delayed, the less my own chance was of going down in the cages – it struck me that the first opportunity to dive in three days was a hell of a time to overhaul a camera which could have been checked out in Durban, or yesterday, or earlier this morning. If divers and cameras weren't ready to enter the water at a moment's notice, the white shark itself could come and go away again, unrecorded.

By midafternoon, when Gimbel and Waterman entered the water, there were still very few sharks; besides, the water was so roiled from the hard blow of the days before that it was decided not to film. But earlier, before the sunlight began to reflect off the ocean surface instead of penetrating it, the visibility had been good enough to identify a school of fish deep beneath the ship as yellowfin tuna, and two lesser rorquals, the "minke" whales of the Norwegians, circled the dead sperm whale and passed several times beneath the bows, white bellies gleaming as they turned on their sides like giant porpoises to inspect the ship. The whales were pierced by the ice-blue meridians that shimmer in the sea at noon, radiating outward near the surface like the fading beams of a light lost in the abyss.

At twilight on that Saturday afternoon, Peter ran me over in the dinghy to W-29; I would rejoin the expedition in north Madagascar late in June. "I'm sorry you're going," Peter said. "Tomorrow's going to be terrific." I knew it would be, and I wished him luck, and jumped aboard W-29, which was picking up her seven whales and heading back for Durban.

"We saved a few hot peppers for you," Willy Christensen said, and Reidar Smedsrud asked, "Was that the New York millionaire?" They had all the newspaper accounts and were full of theories about the "white pointer", which they thought must be what they called "bleu pee-yointer". Captain Haakestad asked if I had gone down in the cage, and when I nodded, he shook psychic pain from one blunt fisherman's hand, a gold tooth gleaming in his smile.

W-29 moved through the darkness, picking up one whale after another by the tiny radio beacons on the marker flags. They were all big sperm, forty to forty-five tons; two exceeded fifty feet in length. Under the seachlights, a number of sharks sliced the black water, and then, from the bridge, a voice bellowed, "Bleu pee-yoin-ter!" I grabbed a spotlight mounted on the bridge and shone it out beyond the whale. Coasting in the yellow ripples of the light's reflection was an enormous shark – fourteen or fifteen feet, all hands agreed. It was by far the biggest blue shark I have ever seen.

The Norwegians were disappointed that their blue pointer was not the fish we sought; they doubted that the white pointer, if there was such a thing, occurred in these waters at all. We winched up the last carcass and started for shore, a hundred miles away. Toward three in the morning, in a sector of strong northeast winds, the whales on the weather side were chafing so badly on their chains that it was decided to flag one and let it go; another boat could pick it up the following day.

At dawn, Captain Haakestad had second thoughts about white pointers. He recalled a monster that had come once to one of his whales, and he also recalled that Arvid Nordengen had actually harpooned an enormous shark only two or three years ago. He was reminded of this when, approaching the Durban breakwater, we met W-I7 on her way out. Haakestad got W-I7 on the radiotelephone, and Nordengen confirmed that he had shot a very big shark of a species unknown to him;

it had weighed at least three thousand pounds. W-17 continued on her course offshore, and W-29 went on into Durban harbor. It was eight in the morning of a lovely Sunday.

The year before, Gimbel had seen photographs of Nordengen's shark, which was unquestionably a white; it had a vertical jaw span of just under three feet. He had also inspected the records of the Durban Shark Fishing Club, which listed a dozen or more white sharks caught each year from the harbor jetties. The director of Durban's aquarium and the director of the shark-netting operation that protects Durban's beaches both testified to the regular appearance of white sharks in these waters, and all this evidence had decided Gimbel on the choice of a South African location. Originally his base of operation was to have been the Great Barrier Reef, but the Reef no longer shelters enough sharks or other large marine creatures to occupy Gimbel's film crew for two weeks. The bleak coast of South Australia had large white sharks but little else, and South Australia was in the temperate zone, where weather and water clarity were uncertain.

There seemed to be little doubt that the white shark occurred off the Natal coast, but Valerie Taylor was bothered by the absence of white-shark jaws and teeth in Durban, even in places where teeth of other species were sold: if six or seven white sharks were sometimes taken in a single month from the jetty alone, as had been reported to Gimbel, and others were taken regularly in the protective nets off the beaches, whatever became of them? "If I were Peter Gimbel," she wrote in her diary for April 13, "I would film one more good feeding frenzy, then head south into colder waters where the seal colonies are situated. I would then kill a seal for burly and set a little fruit tin of whale oil dripping over the stern . . . Ron agrees with me. The water here is too warm for whites and there are no sea creatures suitably large and abundant for a shark as big and hungry as a white shark to feed upon. People here in South Africa who claim to have seen and caught great whites probably wanted to get Gimbel to make his film around Durban . . . I also don't think the crew are as dumb about sharks as Peter tends to think. I hope and Ron hopes he gets his shark off South Africa because that leaves the Australian ones completely for us."

A week later she wrote, "If a great white had been within forty miles of us today, he would have come in. Both Ron and I feel certain of this. They must be very rare in these waters . . ."

But the question of white sharks was already fading in the face of a factor far more serious, an unseasonable bad weather that kept the whale-catchers in port more days than not: between April 13 and May 4, in fact, there were only two days when wind, turbid water or an absence of sharks did not seriously limit the amount of good footage obtained. April 13, the day I left, was indeed a beautiful day, but only a few sharks were attracted to the whale. For the next four days the *Terrier VIII* was held in port by storm, and on the fifth she put to sea but found no action. In this frustrating period, there were petty disputes over such matters as dogs and laundry. As Valerie recorded in her diary: "No wonder the U.S.A. is in such a state if Americans cannot agree on a simple thing like laundry."

The dog problem arose from the presence of a second Norwich terrier, which belonged to Peter Lake, and which Gimbel had banished from the film. "It's really quite touching," Gimbel says, sincerely perplexed, "how crazy Lake is about that awful little dog." Lake no doubt feels the same thing in reverse, because Gimbel is certainly crazy about Billie. To judge from his behavior in its presence, the dog seems to fill a place in his heart that no human being has reached. "I take him wherever I go, even to work," Gimbel says, hugging his dog. "He's just always with me."

"Americans are quite mad over their pets, it seems," notes Valerie.

On April 16, in the evening, the film crew saw the first rushes of the footage taken on April 8. "The sharks looked huge," Valerie wrote. "Stan had a lot of camera movement and so did Jim. Peter G. was very steady. Stan and Jim will have to steady down or people will leave the theatre with jumping eyes after seeing our show . . . I come across okay on film but Ron will have to stop frowning so much."

The next morning the Durban police stopped a car containing Valerie, Peter Lake, the ship's Dutch cook and the cook's girl-of-the-evening, who was "colored"; in South Africa, association between whites and "coloreds" is against the law, and the cops rudely demanded papers and threatened to drag the whole lot off to jail. But as in more civilized countries, the most defenseless person was chosen to pay the penalty for the rest, and the poor tart was taken away.

On April 19, the *Terrier* towed a whale for a long distance in order to attract sharks, and the sharks came, and the water was clear. Ron Taylor killed two white-tips for the cameras, and one of these, in the ecstasy of death, leapt clean out of the sea, to the astonishment of people on the deck. For the first time, big amberjack, vivid gold and blue, came in to feed on the drifting scraps of whale, and this was the day that Gimbel, without warning, left his cage. There was no new kill nor vibrations of distress to excite the sharks, and he had convinced himself that they would not become aggressive with so much ready food available.

"I'm scared of sharks. I've always been scared of sharks and I'm still scared of sharks and I imagine I'll continue to be, because I think that anybody who's not frightened of a shark really is a bit out of his mind," Gimbel wrote later. "It's all very well to say that you have observed a pattern of behavior very closely and that you're armed with a powerhead that can kill them if you hit them with it. All that is on the plus side, but these animals are as well armed as anything alive, and they have a quality that's particularly dreadful that I always think of as the quality of an insane passionless killer. You have no possibility of really anticipating what a shark will do. It's not a mammal and it's a form of life far removed from any mammal; it's an eating machine. Trickery won't work, or craftiness or cunning, as it wouldn't work against an insane mind. Anybody would prefer to stand up to a sane man with a gun because you have some chance of using reason or wile or cunning or craftiness or persuasion on a person with sanity. When you come up against that quality of the passionless killer, you're in bad trouble."

"Peter went out first," Stan Waterman remembers, "and I followed him out; he was always first, in each progressive step we took." Both divers were immediately besieged by curious sharks, but Gimbel's instinct had been correct. At no time on this day or in the weeks that followed, when all four divers left the cages regularly, did any of the hundreds of sharks encountered launch a real attack, though the divers were repeatedly pushed and nudged and bruised; on this first day, a white-tip banged Gimbel's camera hand so hard that it sprained his finger.

The Taylors, unlike Gimbel and Waterman, had friends who had been crippled and even killed by sharks, and that first day they were appalled. Valerie was already in trouble with a shark that appeared out of the murk and entered her cage by the door through which Peter had left: "My hackles rose with horror, and I dropped to the floor." The shark blundered out again. By the time she got reorganized, Gimbel and Waterman were surrounded. "One giant glided in toward Stan mouth agape . . . Stan spun toward the brute just in time. There was no doubt they were risking their lives. I felt the sharks to be only curious but that any small thing could incite their curiosity into frenzied attack."

Typically, Ron was more restrained, but that he acknowledged any peril at all is significant. "I thought they were taking a chance of getting hurt," he told me later, almost unwillingly, adding, "I was apprehensive, yes, but not nearly so much as I would have been with a great white." Subsequently, when Ron himself swam with the sharks, he accepted it with his customary calm. "It was just a job, that's all. Like motor racing. We soon learned that they weren't going to bite us – but still I wasn't the first out of the cage." Ron mentioned that in the open water the divers had to approach each other carefully: once he came up and touched Peter's arm "and he almost had apoplexy".

That evening Valerie wrote, "If this sort of shooting continues, this will be a shark film to end all shark films." Good weather prevailed on April 20, but no whales were found – "a proper washout" – and on April 21 a storm drove them back to port.

April 22. "Today was a fizzer. It just blew and blew and is still blowing . . ."

April 23. "A boring dreary day . . ."

April 24. "Even grayer than Wednesday. I am almost unconscious from boredom. If we don't get some action soon, I shall simply go nutty." That day it rained from dawn to dusk, but the whale-catchers went out on the rain-flattened sea and killed the season's record of thirty-two whales, one of which was rushed by sharks and eaten right down to the spine; after days of bad weather and idleness, this missed opportunity was the worst blow of all.

The next day the *Terrier* put to sea. "It was still a very unpleasant day but Peter G. must be getting a bit desperate, for into the drink we went, rain, swell, and wind unheeded. There was not much action . . . Case's cooking is going to kill me if we don't get some fresh vegetables . . ."

April 26. "Well, Case's cooking nearly killed not me but Philip. We had to rush him to port last night . . ."

Phil Clarkson had been struck down by serious ulcers, which had been worsened by the rough days at sea. He spoke to no one until he became so sick that he had to be taken ashore, where the doctor recommended that he return to the U.S. after leaving the hospital. Phil would not go to the hospital – his compromise was a few days in his hotel room – and he certainly did not go home, but while in South Africa he never went to sea again. His place was taken by Barry Cullen, a diver-yachtsman who performed ably in the following eight days that were to make the sequence on the whaling grounds the best shark footage ever obtained by anybody anywhere.

After dropping off Clarkson, the *Terrier* went straight to sea again, and toward noon, as was usually the case, lay alongside a harpooned whale. The weather and visibility were good but the sharks were mostly rather small, giving the divers a good opportunity to test out Gimbel's theories, which he expounded for me two weeks later in a long letter: "Never fail to react to a bump or a close approach; with a slow-moving or stationary target, they make a close approach or a bump before biting – to see how much resistance is left in the object. Therefore, push them, club them, kill them (with a powerhead) but in no circumstances lie still, play dead, or fail to react aggressively."

Preparations were being made for a night dive: the weather was promising, more sharks were gathering, and the *Terrier*, which was the first to sight the whale pod to which this dead one belonged, had received permission to keep the whale overnight on the condition that she tow it to the slipway the next morning.

It was past midnight before the lights were rigged and other preparations made. Peter and Valerie made the first descent: "As the bubbles cleared," Valerie wrote, "I took my first look. It was my privilege to gaze upon a scene of death and life more horrible and primeval than any I had seen before. Huge sharks swirled in a frenzy . . ."

Peter also mentioned this experience in his letter. He described a huge tiger shark, fourteen to sixteen feet, with a head almost three feet wide, playing with one of the 5000-watt lights perhaps fifty feet below the cages. The monster took the entire lamp in his mouth, reflector and all, then spat it out.

The ocean was still very rough, and with the surge of the ship the electrical lines kept crossing and snarling; at the same time, the cages were banging together and tangling their tethers, until finally the divers' position became precarious. At this point, Valerie wrote, Stuart Cody "leapt into the dinghy, cut the cages free from the ship, and with exceptional skill and courage proceeded to untangle the twisted mess."

In the cages, the divers were not aware that they had been cut adrift. "I wondered why we were unable to see the whale any more," Valerie continued, "and why the light was failing, not realizing we were no longer tethered to anything, but were adrift at night, surrounded by feeding sharks under the Indian Ocean. Not a comforting thought." Then the cages were secured again, and almost immediately a big white-tip got snarled in the moorings, throwing the divers back and forth as it thrashed about. Tangling again, both cages drifted somehow into the red murk at the great cavity in the whale, and meanwhile Peter and Stan were running out of air. Finally, the cages were pulled up tight against the ship's hull, crashing violently against it. "I couldn't vacate the darn thing fast enough. Peter handed out the camera, and then, the perfect gentleman as always, came back down and made his after-you sign. Back on deck, an exhausted trembling wreck, I heard how the cages had nearly been lost when cut adrift. Stuart . . . had the captain maneuver the *Terrier* into position downstream and pick us up. Without his quick thinking we would probably still be going . . ."

Cody was much the most temperamental of the crew, but no one was in any doubt about the contribution he had made from the beginning. Gimbel could have been speaking for all when he remarked one day, "He's brilliant. Sure, he's difficult sometimes, and we tease him about being paranoid or insecure or whatever it is, but he's bailed us out over and over." Even Phil Clarkson, who had plenty of trouble with Stuart over laundry and other nagging matters, acknowledged that Cody – to his frank astonishment – had been one of the willing and effective people on the deck jobs during his own absence.

Toward dawn, the sharks were swarming up onto the carcass in the competition for a bite; the hulk that was dragged ashore next morning was ragged with white shredded wounds. The surface crew accompanied the dead whale from the slipway to the whaling station. There were nine other whales at the slipway, all of them female, Valerie noted, and all of them pregnant. One of them produced a fetus when she was hauled out on the loading ramp – "a small very beautiful little female, perfectly formed. No wonder the whales are becoming scarce. They are not even given a chance to breed. What a curse the human race is. What sins did the innocent and peaceful commit to inherit the likes of us?"

After the first night dive, the *Terrier*'s fortunes changed once more for the worse. A second whale was kept overnight, and a third was towed for many miles to less turbid water, but the wind persisted, the water clarity was terrible, and the sharks did not appear. Between April 27 and May 4, almost nothing was accomplished. Gimbel already had seven thousand feet of exceptional footage on sharks and men, and Lipscomb, having seen the rushes, told him that they had a sequence. But Gimbel felt that this material was less than half what it would be if a whale could be kept overnight and there was fair weather and clear water the next day – an opinion based "not on speculation but on what we had already seen out on the whaling grounds . . . The equipment," he wrote, "was working flawlessly (our only problems had been two Arri magazines that consistently jammed – one had jammed for Ron during the night dive); there wasn't one person in the group I would have traded for any other; and the foul weather couldn't last forever: we were poised. I decided we should hang on, voluntarily go behind schedule, and shoot for the 60–70% of unrealized potential that I knew could be ours with one good break in the weather."

At the end of April the manager of the whaling company gave Gimbel permission to keep a dead whale alongside for a night and a full day, so long as it was taken to the slipway within thirty-six hours after its death, as required by law. Then, at a production cost of five thousand dollars a day, Gimbel sat down to wait for the weather.

On May 1 the ship remained in port, but the following day she sailed. "We are at sea again," Valerie wrote, "our last trip out from Durban after the sperm whales. Poor Peter, the weather is dreadful." May 3 was another barren day of rough seas, gray skies and dirty water, and May 4 dawned the same way. "Jim is nagging Peter a bit about the lack of underwater footage. What he doesn't seem to realize is that one roll of Peter's underwater film is worth more than 50 rolls of his above-water film." One must allow for Valerie's prejudice in favor of the underwater crew, but what she wrote was essentially true: if Gimbel failed to obtain the stuff for which Lipscomb's footage supplied continuity, Jim could take all his excellent material and paper his garage with it.

Compared with Gimbel, however, Lipscomb had little to lose. "We're going to go well over the budget," he told me later, "and that's going to come out of the producer's share. Peter and Stan signed away their fees – they were pretty inexperienced, I guess." (Jim's own salary was guaranteed in the contract; most of Peter's and Stan's had been "deferred until completion of principal photography.") Even worse, as producer of the film, Peter would be held accountable for its failure, and if it came, the failure would be an extremely expensive one which a new film company like Cinema Center would not suffer gladly. As a new producer, with no screen credits to his name, Gimbel could afford it even less. In the film industry, a success (a box-office success, that is – a *succès d'estime* is accounted an abysmal failure) is applauded without reservation, no matter how fortuitous its circumstances or inept the person responsible, and thereafter this person will be awarded enormous sums for film after film, until his ineptitude is unmistakably established by his talent for losing money. But a box-office failure the first time out is almost invariably fatal to the culprit's future as a producer, especially when large sums are involved; extenuating circumstances are as irrelevant as ignorance of the law, and Gimbel knew it. "I was frightened," he wrote me on May 13, "not of the Great White, not of the other sharks we saw and swam with all the time, but of failing – the only thing that has ever frightened me about this project on a long-term basis."

DURING THE MORNING of May 4 the weather rapidly improved, and toward noon the *Terrier* came up alongside a medium bull sperm whale killed by Arvid Nordengen on W-17. Already sharks had begun to gather. The surface was calm and the sky and water reasonably clear, though a big swell was still rolling by from the storms of the weeks before.

Stan Waterman and Peter have since told me that, purely as a spectacle, my own day in the shark cage on April 8 was the most spectacular of all before May 5, but as Peter says, the difference between April 8 and May 5, when he won his gamble, was not a matter of degree but a quantum jump; that this is no exaggeration is evidenced by the fact that more underwater film was shot in thirty hours than had been shot in the previous thirty days.

The first day – May 4 – was quiet enough; in fact, Gimbel dismisses it in a few lines: "The sharks did not gather quickly, but there were enough around – perhaps twenty to thirty – to give us the atmosphere we wanted for certain connecting shots." (Bob Young, the director, who had seen some of the rushes in New York, had cabled that the high points were marvelous but that there was no connecting tissue in between. ) Stan and Valerie, who also took notes on these thirty hours, were less casual about May 4. Both mention the sharks' lack of interest in the whale and close attention to the divers, who were kept so busy fending them off that they failed to obtain the connecting shots they needed. Using her bang stick like a club, as all four had learned to do, Valerie "thumped and whacked as fast as I could . . . Part of Peter's plan was to film the sharks feeding on the whale but we never made it . . . the sharks beat us back every time. Ron finally killed one that came too close. It died a beautiful death, skidding in ever-decreasing circles . . ." Perhaps because so much meat was available, the surviving sharks paid no attention to the dying one, on this occasion or any other.

Meanwhile, more film magazines were jamming, and Peter became upset and depressed. When camera difficulties were compounded, after dark, by a prolonged malfunction of the lighting system, he lost heart entirely. Lipscomb was still perplexed by this two months later. "Here Peter had the whole thing in his grasp, and he just quit. It was a beautiful clear still night, with a full moon, and Tom Chapin was singing folk songs up on the bow, and everybody was ready, and the lights were finally working, and after everything he had gone through, Peter chose this moment to get discouraged. If Stan and I hadn't really kept after him, he would have gone to bed."

"I was feeling exhausted and depressed," Peter wrote me, "unable to shake a feeling of irritability and the sense of a wasted opportunity from the . . . afternoon, when I had had two magazines jam in rapid succession . . . Nevertheless, Waterman and the Taylors were eager and ready and Lipscomb was pushing me, all of which got my fight or my pride – or something – stirring, and we got into the water about 1.00 a.m., May 5. Both cages had two cameras in them: Stan and Ron with two Arries, one with a 9.8-mm lens, the other with an 18-mm; and Valerie in my cage with an Arri mounted with an 18-mm and the Aquaflex (Eclair) with a 35-mm."

In the still night, the wet snuffling of the feeding, the moist slap of meat, was very audible. Sharks are said to be nocturnal feeders – they are equipped with a topetum membrane, also found in certain nocturnal mammals, that reflects and amplifies ambient light – and all the wildest feeding frenzies observed from the *Terrier* occurred in darkness. But at night they appear to see less clearly, for they blundered into the cages much more often and with greater impact. To judge from the prevalence of shark attacks that occur in turbid water, a person's chances are improved when the shark can see him clearly: add to this man's fear of darkness and the ever-present possibility that the lights might fail when the divers were out in the open water, and the nervousness of Gimbel's companions becomes very understandable indeed.

"Peter might have swum out at night," Stan Waterman said, looking worried. "We followed him out in the daylight, but whether we would have done it at night is a moot point." Stan was relieved to be in the cages during the night dives, after the nervous strain of swimming with the sharks during the day. Valerie dreaded above all the failure of the lights, and she spoke sharply to Peter in advance. Gimbel agreed that he would not leave the cage; he climbed out the top hatch instead and sat on one of the static buoyancy tanks to give himself a clear field for filming. "This was the scene," he wrote. "Scores, perhaps even hundreds, of sharks, almost all enormous . . . swimming not quite lazily but not lashing about, crisscrossing in a stack down to about 40 feet. The sea, illuminated only by the big lamps, contained no trace of blue. There was a remarkable feeling of being within a translucent substance of varying tones of gray – from almost white to light medium – then, at the perimeter, suddenly black, blank (but I hardly ever looked there – I'm too savvy: I know where panic waits). You must remember we were illuminating the sea from *within*, and one never sees that by day. Besides, we were 9000 feet above the bottom – i.e., no bottom. It was surely the most surrealistic sight of my life: we were encapsulated in molten rock salt, a red-pink-white wound the size of an open rowboat pulsated 10 feet away in rhythm with the swell, and the sharks swirled about and drove into the gnawed-out hollow in the whale, latching on and vibrating like outsized tuning forks gone mad until the gobbet between their jaws would tear loose and they would swim away with a plume of pinkish blood at either corner of their mouths."

The divers' estimates of the shark numbers do not vary much, though Peter is an optimist and, as a rule, his figures and Valerie's tend to be highest, while Ron's are invariably lowest. "Hundreds of huge sharks crisscrossed in a steady moving pattern," Valerie wrote in her diary. "Several long, beautiful blue sharks glided between the heavier, uglier white-tips; one or two fat lazy greys completed the picture . . . What peace and tranquillity above, what hell and carnage below." According to Stan, the shark numbers were hard to estimate, but between a hundred and fifty and two hundred would be an intelligent guess; he too noticed the big blue sharks, which he estimated at about fourteen feet, and several white-tips perhaps twelve feet long that must have exceeded a half ton. Ron Taylor thought about it briefly before speaking: "In excess of a hundred, I should say." Knowing Ron, it is safe to assert that a hundred sharks was the very minimum.

The pattern of shark species remained constant. The great majority, day after day, were white-tips, but a few blues were always scattered among them, as well as *C. obscurus*, known as the dusky shark or "lazy gray". Occasionally a tiger shark appeared, and perhaps a dozen times the divers saw the strange shark with the rolled dorsal that we had first seen on April 8. In most respects this creature resembled the dusky, but whether it was an aberrant form of *C. obscurus* or a new species, nobody could say: the shark authorities at the museum in Durban know nothing about it.

Peter's letter continues:

"The sharks were feeding on the whale ferociously and brushing close by me and from time to time bumping me roughly. I looked everywhere for the Great White – as I always do when my eye is not on the viewfinder . . . I would have had no particular difficulty in leaving the cage and swimming among the sharks. I think only lethargy, a leeriness of all the lines and electrical cables in the water and an overwhelming determination not to have anything go wrong before the next day (when I was sure the jackpot awaited us) prevented me from doing it. I regret now that I didn't; it would have made all the difference in the world to our night sequence. The thing that will really hit an audience like a ton of bricks watching our film is not sharks but people out in the open, surrounded by sharks, brushed by sharks and, above all, swimming with them on their own terms: vulnerable . . .

"Before I went to bed – at about 4.00 a.m. – Jan Moen said he wasn't sure there would be anything left of the whale by morning. So Tom Chapin fired up some of Jim's big deck lights, which Moen felt might slow the sharks' eating down. Maybe it did – who knows? By the light of those lamps Jim shot a roll of the sharks hard at work, and they didn't seem concerned about the illumination!

"At dawn, May 5th, the sea was even calmer – long, easy ocean swells laboriously heaving under an oily smooth surface. As I peered out the port, it became plain that, in fact, the sea *was* oily: a heavy slick of blackened clotted blood, pieces of blubber, and oily exudate surrounded the ship. I ran out on the deck with despair welling up in me, sure the whole damned whale had been eaten down to the spine in the couple of hours since I had last seen it. The carcass was still floating off the bow of the *Terrier*, badly damaged but no more than 15 to 20% eaten. I was astonished because, as we had all heard, the Company has had five or six sperms completely eaten down to the backbone this year within a 6 or 7 hour period. I tried to extrapolate – to imagine what kind of feeding performance could accomplish such a thing if the ferocity and numbers of sharks we had seen in the past 8 or 10 hours could only – Wow! *Only!* – do this. The sharks were still at it, but in fewer numbers, at least so it appeared from the surface, and certainly less aggressively. I went below, gave up my theorizing, and ate my own measly breakfast . . .

"We were in the water by 9.30 – Stan, Ron and I each with a camera, and Valerie with a powerhead riding shotgun. The three of us each had a powerhead looped around our wrists too – just in case of the sudden appearance of Big Whitey, or some other crisis. As we left the cages, which we did immediately, wishing to shoot footage of the group near the whale surrounded by sharks, we came under heavy pressure at once. I realized we had made a tactical error in not having two divers riding shotgun and two filming instead of one and three. But I was unwilling to change plans, being very greedy to roll the maximum film through the cameras and realizing, also, that it would be more exciting this way. Fifteen or twenty big heavy white-tips and duskies, ignoring the whale, closed in on us, herding us into a tight little bunch. I became totally conscious of the sharks and was unable to hold my takes nearly as long as I knew the shots demanded. Valerie was moving gracefully around the perimeter swatting sharks, but we were being continually brushed and bumped from all directions. For a few minutes I felt that we were pinned down, unable to move closer to the whale or back to the cages. And to make matters worse, I knew I wasn't shooting well: unable to concentrate and lacking the will to pull back out of our herded-in knot to get the overall action I wanted. Our film ran out and the pressure eased almost simultaneously, or so it seemed.

"It's difficult to be objective. In retrospect, I feel we were no more threatened – very likely less – than we were later in the day; but I *felt* the pressure; I *was* alarmed. I think it may be analogous to the first ski run of the day, especially if you pick a challenging trail that turns out to be a little icy: some things call for a warm-up. I shot that first roll cold, and suffered through it. During the rest of the day we exposed ourselves even more boldly, but I was filled with a sense of controlled excitement, exhilaration and dominance – never anxiety. I am sure we all felt the same way."

This was true. As Waterman noted, ". . . as often as we were bumped and jostled that day, no shark ever proceeded to the bite stage. But the contacts were so frequent that we became inured to them. A wild fatalism overtook us . . . we felt a growing sense of immunity . . ."

For almost six hours, from just after nine until just before three, the divers only left the water to reload cameras and obtain fresh air tanks. For twenty-four hours, the blood and smell of death had been drifting down the current, which the sharks followed upstream like a path. The whale's viscera had been opened up, and the blubber was slashed through to the meat, and meanwhile the vibrations of hundreds of big avid bodies drew more and more shadows from the sea. The divers swam in a maelstrom of big sharks, and not once was anyone attacked or even seriously threatened; there is always a first time, and they knew it, but it never came. Waterman felt that the sharks, while curious, accepted man as another scavenger. In the gathering of albatross and jacks and sharks, with the attendant pilotfish and remoras, the strange bubbling black creatures were of no significance.

"Because of a shortness of magazines – two had had to be taken out of action for repairs – and the speed with which we were shooting, the surface crew couldn't keep us evenly supplied with loaded cameras. So we intentionally permitted ourselves to get out of phase with one another. That is, anyone who finished a magazine would hand his camera into the dinghy, get in himself and go back to the *Terrier* for a reload. It's interesting that we'd always return to a cage to board the dinghy. I think we all felt cautious about having our legs dangling like sausages into the water while hauling into the boat . . .

"When only three of us were in the water Lake would come in to take stills. Because of the situations I've just described he got plenty of chance. As you well know, sometimes quite hazardous situations can be funny: often when I looked at Lake, I would sight him in near trouble with a shark moving in very close and him flicking or pushing at it with his little camera. The sight had about it that unreal feeling, when . . .

the passage of a single second will cue a laugh or a moan. Luckily, Peter always came up with laughs."

For all his experience with such perilous pursuits as auto racing and parachuting, Peter Lake was not yet a confident diver, as he himself was the first to admit, and the expedition had been on the whale grounds for weeks before he brought himself to put a tank on and enter the shark cage. As the still photographer, he felt obliged to do so, quite apart from the galling knowledge that he had been missing extraordinary material. But he had only been underwater a few times by May 5, when, in addition to the sharks, he had an experience that fortunately he did not understand while it was happening. The line securing Valerie's cage to the whale chafed and parted, and the cage, which was not stabilized, descended rapidly to eighty feet before Valerie realized that the first air tank was exhausted and got the valves switched to draw air from the second. To Valerie's relief, the cage rushed to the surface, where she discovered that Lake thought she was simply giving him a joy ride. In any case, after May 5, Peter was a certified veteran, and has dived with great confidence ever since.

"As midday approached, the vigor of the feeding ebbed," Gimbel goes on. "This was a tendency we had seen on other days as well. I don't mean that the action got slack; it was just distinctly less wild and the sharks were more dispersed.

"By about 2 o'clock things were becoming quite lively again. During the course of the day about 50 albatrosses had gathered around the carcass and any time you liked you could see them pecking at the whale . . . often with their heads only inches away from the jaws of the feeding sharks. Looking up toward the surface, their fat, round bellies looked, as Lake put it, like toilet bowls. They are truly enormous birds. From our rather unusual point of view, they impressed us as utterly revolting. Once I popped my head out of the top hatch of a cage . . . and found myself staring into the beady and fearless eyes of one of those birds . . . Quite surprising. They have a beak about 6 inches long, you know, with a terribly pointed, bony triangular cap at the end of it."

Allegedly, Ron was more nervous about the beaks of the albatross than about the sharks. "Why, those things could *hurt* you!" he exclaimed at one point, in genuine consternation.

"A bit past 2 o'clock," Peter continues, "I was aboard the *Terrier* getting a fresh air tank and new film load. Valerie was either already on the ship or arrived shortly after me; I forget which. In any case, she was plainly exhausted and I told her she should not rejoin us until she was rested. I didn't expect to see her in the water any more that day because she wasn't simply winded; she was worn out.

"The dinghy took me out to the whale and I entered the top hatch of a cage which was at the surface, unoccupied. The other cage, with Ron and Stan in it, was about 20 feet down on the opposite side of the whale. It was about 3 o'clock and I knew this would be my last magazine. I checked the camera settings, looked around for the Great White and opened the side door. The sharks were massing again, as we had always observed them to begin doing at midafternoon. (Don't forget, we were at 30° south latitude, approaching the winter solstice, and the ambient light underwater begins to fade rapidly after 3 o'clock.) I hesitated in the open door for 2 or 3 seconds, full of strangely mixed emotions: first I felt utterly calm – almost as if I had suddenly been graced with invulnerability – then blissfully happy, luxuriating in the realization that we had actually pulled it off, that no matter what happened to that final roll, we already had in the can footage that was far beyond anything of its kind ever attempted, let alone achieved; then, I felt a kind of self-conscious realization that neither of the first two feelings were nutty – that they were, in fact, quite sound, the latter simply an empirical truth and the former the hard-earned reward for pushing day after day in the face of our uneasiness and our fear to find out for ourselves precisely where the limits were, just how far we could go, how openly expose ourselves in swimming about among these sharks *in these maximum conditions* – i.e., bloody water, active feeding . . .

"I swung out of the door, swam to the opposite side of the dead whale where the greatest activity was, and began shooting. Stan left the cage below, carrying a powerhead, and joined me . . . Ron was sitting on the top of the submerged cage and filming from there. Then Valerie appeared. I must have been filming when she joined us because I never saw her arrive. Suddenly, I looked up and there she was, rapping the sharks about with Stan. I knew how tired she was, hadn't expected her back in the water, and I was touched to see her there . . .

"I moved in very close to the whale, maybe 3 or 4 feet from the heads of sharks that were clamped on and shaking out mouthfuls of flesh. Security at my back gave me no concern, with Valerie and Stan there, but at times the surge would carry me so near a vibrating shark that I was worried one of them would hit me accidentally as he came away from the whale. Even moving slowly they hit with great force – 500 lb of mass, and the front margin of the snout is hard and solid. Once their bite was torn loose they would swing their heads in an arc – this was the move that worried me – and swim away chewing. But clearly it was a day without reason to worry.

"The last magazine ran out. It was the 18th one; we could have made it 20 – there was plenty of daylight – but everyone was tired and I knew we had the material with a fair margin to spare."

For Valerie and Stan, at least, it had not been "a day without worry". While trying to guard Peter, Valerie had a bad experience when a feeding shark, shuddering with effort, turned up beside her. "His vibrating body next to my back felt terrible," she noted in her diary, 'like some primitive shuddering nightmare." As for Stan, he considered the free swimming among the sharks "extremely dangerous", and his own worst moments – unlike Valerie's dread of the dark and the unknown – usually came about when he felt himself cut off from the other divers: he feels certain that the sharks, with instincts tuned by millions of years as successful predators, can sense the weakness and vulnerability of the single individual cut off from his group. Finally, it seems evident that a man taking extreme chances is also endangering his companions, none of whom was pleased to see Gimbel, on that final dive, swim right into the whale's wound among the sharks. Waterman deplored his "bold and frightening actions" and Valerie wrote crossly, "Peter Gimbel seemed to like it up there in the gore and blood surrounded by sharks, he took so darn long getting those last few shots . . ."

Subsequently Valerie told me that Ron, too, had kept a diary. The news surprised me, and I asked her what he had thought about May 5. "I don't know," Valerie said. "I haven't read it." She shrugged her shoulders, looking wistful. "I don't know Ron," she said, after a moment. "We get on fine, but I don't really know what he wants, or what he expects of me. I don't even know what he likes about me, or even if he *likes* me!" She laughed, shaking her head. "I don't know him at all."

When I asked Ron about his diary, he looked very uncomfortable. "Well, it's not really a diary," he said. "It's only got about thirty words in it." He acknowledged dutifully that May 5 had been "incredible".

In the long days of bad weather before May 5, Ron had occupied himself in part by honing both edges of his diver's knife to a fine edge. Previously he had had poor luck in killing these big sharks with the bang sticks – the ammunition simply was not strong enough – and on May 5, when his camera ran out of film, he busied himself, as Valerie said, in "slitting open every shark in knifing distance". Because the flanks were too tough to penetrate effectively, he evolved a technique of stabbing the sharks in the throat as they swam by and letting the knife pass down between the pectorals to lay open the belly. "Ron hates sharks," Valerie has written. "So do I. Many of our diving friends in Australia have been mauled and sometimes killed by these vicious monsters. Ron wanted the personal satisfaction of killing sharks with a knife. He soon discovered the only place the knife would penetrate, so concentrated on throats and bellies with dazzling success."

At times Ron's calm seemed so unusual that the last paragraph of Valerie's diary for May 5, reminding us of the stress he shared with all the rest, comes as a distinct relief:

"I crashed into bed immediately following dinner, an exhausted mess; only sleep eluded me. At 9.30 p.m. I was still fighting in my mind. A thousand sharks plagued me and gave me no rest. Ron was tossing around like he had a shark in his bunk with him. I gave him a call and we both took a tranquilizer. Finally we slept."

ON MAY 18 the film crew left South Africa for Ceylon, flying by way of Nairobi and Bombay; the *Terrier* had departed nine days earlier. In Colombo the crew was met by Rodney Jonklaas, a Ceylonese diver and exporter of tropical fish whose enthusiastic report had encouraged Gimbel to commit a large segment of his shooting schedule to the exploration of the wreck of the *Hermes*, a 450-foot British aircraft carrier bombed and sunk by the Japanese off the Ceylon coast in 1942. Enormous sea bass and big shark had been reported around the *Hermes*, which was readily accessible on the coastal shelf, in 175 feet of water, and there was no reason why the great white shark should not turn up there, though no one thought this very likely. The *Terrier* would be anchored to the wreck, and the film crew would live onshore, in the nearby coastal town of Batticaloa.

From Colombo, the crew drove overland to Batticaloa, stopping off at Kandy, the "City of the Kings"; Kandy was the first thing – and for most of the crew, the last – that everyone liked about Ceylon. They left Kandy in good spirits, taking with them a large cargo of Oriental firecrackers, some of which were fired off on the road to Batticaloa. There they moved into the Hotel Orient, the only real building in town, and exchanged stares in the humid heat with the Tamil people of the coast, who impressed them less favorably by far than the inhabitants of Kandy. Few of the crew had experienced the swarming squalor of the tropic shores, and nobody but Valerie, a fanatical tourist, was impressed by what he saw.

On May 28 the *Terrier VIII* put into Colombo with a faulty generator; she came around the south point to Batticaloa on May 31. Work began the following day. In addition to Jonklaas, the divers had been joined by Frank Mackey, a diver from Singapore and a "very handy guy", according to Clarkson, especially while Waterman was sick with a cold. Waterman was one of several people who got sick in Ceylon, and sickness seems symptomatic of the whole Ceylon experience, which went badly from the very first.

At the *Hermes* a strong unpredictable current that could not be related to the tides was a serious complication, and the water itself was turbid with plankton, limiting visibility to a dim fifty-feet maximum. The *Hermes*, which was never visible from end to end, was little more than an olive-brown amorphous mass on a dead bottom; underwater lights revealed that the hull was covered with rust-colored cup corals, with a scattering of white Gorgonians. On the foredeck was a Bofors gun which, because the ship was on her side, pointed straight up toward the surface.

Since they were working in much deeper water than off Durban, the divers wore twin tanks, but still their "bottom time" was limited to a maximum of twenty minutes. In the depths the compressed air was burned up faster, and on the way up, the divers had to pause for fifteen minutes or more at depths of twenty to ten feet, clinging to lines suspended from the surface; here emergency tanks had been hung down for divers who lacked enough air in their tanks to decompress. Decompression was necessary to avoid the "bends", a very painful condition brought on by the formation and expansion of nitrogen bubbles. The nitrogen, absorbed from the diver's tank under conditions of great pressure, comes out of solution from the body fluids much as compressed gas is released from a newly opened bottle of soda water, and if the diver returns straight to the surface, where the air pressure is negligible, the nitrogen bubbles, accumulating too fast, may literally burst the tissues as they expand. Should the nerves be damaged, paralysis occurs, and sometimes death.

In these difficult circumstances, each man was limited to two short dives a day, and a whole week was required to shackle the *Terrier* to the wreck and rig the light cables and "down lines" that guided the divers and tethered the cages in the strong current; the cages – secured to the line by a ring – had to be towed down by the divers. Also, the current kept fouling the lines, which would chafe and part on the rusty skeleton of the rotting hull. On June 7, the bow anchor chain let go when the girder that it encircled pulled apart with the ship's surge; four

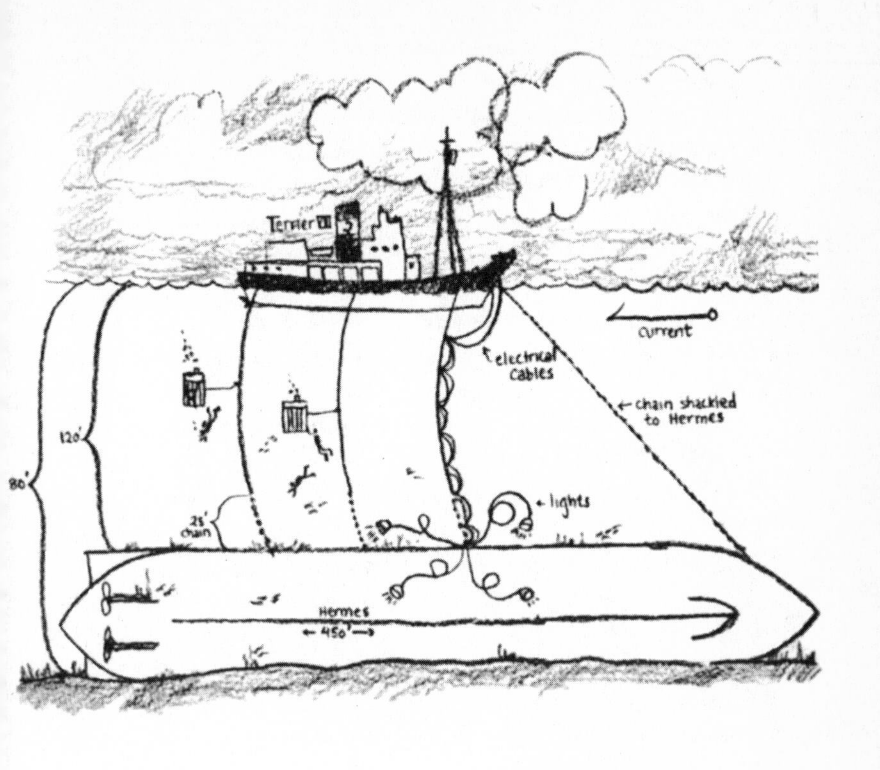
Terrier XII
current
electrical cables
chain shackled to Hermes
lights
120'
180'
25' chain
Hermes
← 450' →

hours were required to secure it once again to a heavy gun turret. A week later, a light cable left overnight was parted by the ship's swing. Meanwhile, in the hot and windless days in the lee of Ceylon, the ship's deck roasted.

"Visibility has been appalling," Valerie wrote on June 5, "and Peter G. is getting very worried. It looks like it could be a washout . . . We have seen no sharks. Peter G. still seems to imagine that he is going to get the great white. For such an intelligent man, he sometimes surprises me. What would a shark that prefers cold water and feeds on seals be doing in these warm seal-less waters? I feel Peter is pushing his luck trying for whites here. If he gets one, it will be a miracle. Of course there are the usual stories going around, but absolutely no proof at all."

In Valerie's opinion, the whole film could have been shot off the coast of Australia. "Ron and I told him that beforehand, but he wouldn't listen; he had his mind set on South Africa." Valerie is intensely patriotic, and no doubt feels privately that all white sharks are better dealt with in the territorial waters of Australia, where they belong.

The "usual stories" Valerie refers to were based on accounts given to Rodney Jonklaas by the local Tamil fishermen, who claimed that a man-eating shark of over three thousand pounds had once been taken there. Jim Lipscomb, who had gone off to film the local tuna fishery while the *Terrier* was being secured to the wreck, was also told of a "devil shark" whose description appeared to match that of the great white. Gimbel had no serious hope that the great white would turn up at the *Hermes*, but its appearance there was at least possible. Many great whites have been taken in tropical waters, and their diet is scarcely restricted to seals; one of the white sharks taken off Montauk in 1960 had a bellyful of periwinkles.

On the 8th of June, some underwater filming was attempted, but good photographic range never exceeded fifteen feet, and visibility beyond twenty-five feet diminished swiftly to the vanishing point. The cameramen needed constant help from four supporting divers, two managing the lamps and two who did nothing but clear the lines and cables. Stuart Cody, fearing for the safety of the divers with so much electricity in the water, had rigged his lighting system to short out if the slightest thing went wrong; since small emergencies were constant, the system blacked out continually, and the sudden darkness, especially for the less experienced divers like Lake and Clarkson, added a very unpleasant dimension to the whole experience. Lake found the *Hermes* diving far more disagreeable than swimming with the sharks; later he called it, with great feeling, "a real horror show".

Worst of all, there was nothing to film. The wreck abounded with small fishes – batfish, angelfish, scorpion fish, lionfish, small barracuda – but a moderate-sized sea bass, glimpsed by Gimbel in the shadows of the propeller shaft, was the first and last of this large species that was seen.

Peter was already worried:

"Must write briefly: tired, hot, discouraged . . .

"Water clarity on *Hermes* frightful. None of the giant sea bass seen last year. We are shackled to the wreck, down lines for cages rigged and tested, also down line for light cable, but with this turbidity – plankton, actually – we're dead.

"Badly hit by bends last Sunday along with Waterman and Taylor; they were touched only lightly. I knew I was caught after about 5 min. decompression at 10 feet. After 10 min. I was in acute pain, right shoulder radiating to elbow & wrist, and went to surface to tell Valerie, swimming above, to set chamber up. That ten-foot ascent brought on pain so intense that when I opened my mouth to yell at her, I was astonished to hear a voice emerge. It felt as if a colossal hypodermic needle had been driven into the shoulder joint and was introducing pressure enough to explode the arm and shoulder. We were operating *well within* the safe limits of standard decompression tables. My theory is that we were dehydrated and the so-called slow tissues were not being bathed by a large enough reservoir of fluid to absorb normally the nitrogen from them. Also, in my case I had descended very rapidly, giving me perhaps one extra minute at depth, and I had been working on a shackle about 6′ deeper than Ron and Stan. Then, too, all the scar tissue in my shoulders can't have helped. No lasting effects, and shoulder perfect now . . ."

There is a new theory that a rapid descent into deep water may be more critical in an attack of the bends than the rate of ascent; Gimbel, who was not acquainted with this theory, always descended very rapidly.

When I rejoined the film crew three weeks later, Valerie told me that she had been much more frightened by Gimbel's attack of the bends than by the shark marathon off Durban. That day she was acting as safety diver, watching from the surface to make sure that all was well as the others came up to decompress. "Then suddenly Peter popped up right beside me. He looked awful and sounded awful, yelling at me in that horrible voice – I could see straight off he was in terrible pain." Valerie swam to the ship to get help, but Phil Clarkson was on his way ashore – his son Lindsay was sent off after him in the dinghy – and Lipscomb and Chapin were busy recording the emergency (as Gimbel would have wished), and there was no one else in sight.

With a crewman, Valerie gazed helplessly at the recompression chamber, which nobody but Gimbel, Clarkson and Cody knew how to work. Then Jan Moen turned up and, for want of a better plan, they hosed the chamber down, to cool it. By this time Cody had been found and he soon had the chamber ready to go, but Peter had decided he was better off right where he was – it was simply too painful to go to the surface. Ron Taylor, though much less hard hit, stayed with Gimbel at the decompression level. During the two and a half hours that the divers remained underwater, using up one tank after another, Valerie watched over them from the surface. She was stung badly by small jellyfish, and meanwhile felt sicker and sicker as the sea bobbed her up and down, and finally she vomited. Since Stan Waterman still suffered from a cold, he was forced to descend slowly; he was "bent" only slightly, and came out of the water after decompressing for thirty minutes. A few years ago, while shooting an underwater film outside the barrier reef of Bora Bora in the Society Islands, he had run out of air at 130 feet and had to come up too rapidly. He was "bent" immediately, and for thirty-six hours, until he was finally flown to a recompression chamber aboard a French aircraft carrier that answered a radio call and met Waterman's seaplane at Mururoa, eight hundred miles east of Tahiti, was so paralyzed that he could not even urinate. To this day, Stan has very little sensation in the entire left side of his body.

Once out of the water, Peter drank off two vodkas to ease the residual pain. After supper he went right to sleep, and the next morning he felt fine. In fact, he was ready to go down again, but Clarkson, supported for once by the others, refused to let him dive until he'd had a day of rest.

Two days later Valerie wrote in her diary, "This diving on the *Hermes* seems to be jinxed. Everyone seems to be falling apart." Waterman was out with a cold, Gimbel was still bothered by mild bends, Lake had ear trouble, and several others were recovering from or coming down with something. Then one of the firemen was sent off to the hospital with what he claimed was appendicitis; his affliction was later diagnosed as gonorrhea.

That same evening Claus, the nineteen-year-old cabin boy, collapsed in a stupor that bordered on unconsciousness. The local hospital diagnosed his case as "sinusitis", but it seemed more likely that his problem was extreme misery brought on by the persecution of the Dutch cook, a varlet whose ill repute among his mates was growing daily. Common homesickness was also suggested, for Claus had never been away before. Since he failed to respond to any treatment, he was sent home a few days later, accompanied by another crewman.

This man, Francis deNikker, was scheduled to appear before a board in Cape Town that would decide whether he was white or colored; he was willing to spend his entire savings on the trip in the hope of being pronounced white. Francis carried letters from Jan Moen and from Gimbel which said they thought of Francis as a white man and would entertain him at their table, etc., but his friend Sam Lloyd did not think he had much of a chance. "Cape coloreds" like Francis deNikker are the worst victims of apartheid: at least the Asians and Africans know who they are.

The Francis crisis elicited a social consciousness among the film crew which heretofore had been quiescent. Cody and others had resented the bad racist jokes with which their South African dates had sought to regale them, and Lipscomb and Chapin remembered a day when an excursion out of Durban had foundered because the Indian friend they had brought along was not allowed anywhere they wished to go. But these episodes seemed less immediate than the tragedy of Francis, whose interest, hard work and loyalty had won the respect of all.

Jan Moen has lived in South Africa for fourteen years. He is married to a South African girl and likes the country, but in his gentle way he makes it plain that he does not like apartheid. "It seems funny," he says, "that two brothers might be classified in different races, and not be allowed to associate. Or a man might one day be forbidden to associate with his wife and children." Often "color" has less influence on a man's classification than the people he associates with, and a story that Jan told me points up the essential lunacy of the system: two "Cape coloreds" on the ship are so "white" that they often go to European bars and are not questioned, but one night they were accompanied by a second engineer, who looked so "colored" that he was told to leave. The second engineer was a "white", but had forgotten the card that proved it. In other words, this weird classification of human beings isn't even efficient, so that in the end all the dull cruelty it has meant for millions will have been in vain.

Claus's place as messboy was taken by Lindsay Clarkson, Phil's fifteen-year-old son, who shortly thereafter got third-degree burns on his left hand while setting off some odd Oriental rockets. Gimbel found him in the medicine locker, struggling to doctor himself. "I got kind of a bad burn," said Lindsay, extending his charred flesh. "He's a teeth-gritter," said his father, who is a teeth-gritter himself. For weeks afterward Lindsay carried before him a terrible white-anointed hand, like a reproach. Nevertheless, he stuck gamely to his new post, and continued his duties on the return voyage from Colombo to Diégo-Suarez.

On the 12th of June, persons unknown stole a camera from the Taylors' cabin and an electric razor from the cook. The police of Batticaloa, Ceylon, have the same instincts as those of other climes, and though the real suspect had disappeared, they got a quick confession by beating a whole string of Tamil fishermen, putting two in the hospital. This event added local hostility to the troubles of the expedition to the *Hermes*.

Despite all these calamities, Gimbel persisted for another week. The ship had come forty-five hundred miles from Durban to Ceylon, at a cost of eight hundred to nine hundred dollars a day, and he still had hopes of underwater footage that might support Lipscomb's fine surface shots of local fishermen, outrigger canoes and cast-nets. But the water conditions never improved, nor did the lighting ever achieve a proper balance between safety and efficiency, and small accidents kept occurring. One day, due to water pressure on a blocked sinus, Clarkson startled his diving companions by popping up with a face mask full of blood.

Phil, who was living and eating ashore because of his ulcers, was almost his old self again – "Phil Clarkson is only a little guy," Valerie noted on June 13, "but he sure is handy on the water" – and was working so hard as a diver that twice he ran out of air. On one of these occasions, he just barely made the emergency tank suspended from the surface. Still, his illness had dampened his spirits, or so it seemed, for a growing pessimism was very noticeable.

One day Gimbel, desperate for action, shot a whole roll on negligible material before he discovered that his camera had not been loaded. One winces to think about the film crew's morale if this had been the day that they were waiting for. "Every cameraman is responsible for his own camera, of course," Peter admits, "but Ron told me it was loaded, and I was in a hurry, and he is the one person on this ship whose work I have never had to check. He's conscientious, willing, and effective. In fact," Peter said, after a moment – he did not say this lightly – "he's by far the best diver I've ever worked with. Plenty of people dive as well, I suppose, but Ron's also superb in his preparations, in care and maintenance of equipment, and in foresight: he anticipates trouble. He sees what has to be done and does it while other people are still asking how they can be of help. This was the exception that proves the rule: Ron just doesn't make mistakes.

"Another thing: Ron nearly always came up with sound solutions to problems. He has the instincts of a problem-solver. He doesn't waste time admiring the merits of a scheme or a piece of equipment; he probes for the flaws, the points where an operation might be most likely to fail, the intrinsically weak components of a piece of equipment. He is justifiably proud of his technical ability and of his ingenuity in applying it to diving. He is a real pro who retains the enthusiasm of an amateur – which I reckon is about the highest praise you can hand a professional."

On June 14 the ship's crew enlivened a hot dull day with the capture of a brown-and-yellow sea snake with ornate white markings on its tail. Sea snakes are true reptiles of the family Hydrophidae, and their poison is related to the nerve toxin of the cobras. Though ordinarily shy and slow to anger, they become excitable and vicious when hurt or provoked – so much so that Lloyd's of London, while perfectly willing to insure the Taylors against shark attack or drowning, will not insure against death caused by sea snakes, which infest those waters of northern Australia also infested by the Taylors. "Why, they'd be paying out on sea snakes all the time!" Valerie says. Weeks later, she was still annoyed at Rodney Jonklaas, who had grasped the captive snake behind the head and played with it until it became annoyed, then tossed it overboard. What bothered Valerie was that the divers were to enter the water an hour later. The incident didn't bother Ron, who knew that even a very angry snake would not recall what made it mad an hour later, much less lurk about to take revenge. But Valerie cited episodes of vexed sea snakes sinking their fangs into the boats of their tormentors: she refused to be mollified.

On June 16, inexplicably, the current slackened for twenty-four hours. That night some eight-foot stingrays were located by the cameras, flapping like ghosts over the ruined vessel. But the visibility remained poor, and except for the rays, there was little to film. Even the deadly sea snakes that swam around the ship at night could not be located underwater, and as for sharks, not one had been seen in eight days of diving. Indeed, the use of shark cages had been discontinued on June 12, despite continuing reports of the presence of the "devil shark" only a few miles further offshore.

Before the last night dive, Valerie seemed especially nervous; sitting down beside Ron on a bitt, she took his arm and told him to stay close to her. "I want to be able to watch out for you," Valerie said. "I couldn't stand it if anything happened to you." To the others she said, "I'd find it very difficult to replace him, you know." She smiled as she said this, but nobody took it as a joke, least of all Ron, who disliked the suggestion that he needed to be watched out for. Putting his hand on her shoulder, he said quietly, "Now, Valerie, you just take care of your*self*, all right?"

Though she liked Ceylon, Valerie's morale was noticeably affected by the long hot weeks of frustration. In her diary, for example, she deplores Peter Lake's diving technique and underwater manners; his flippers, she felt, were forever in her face. Annoyance with Lake's raffish ways crops up here and there from the start of the diary, though later it is replaced by real affection; one suspects that she had not yet forgiven him for his free use of four-letter words, which Ron didn't care for either.

"I have heard Ron curse just once," Gimbel told me. "It wasn't very spectacular. He said, 'God damn it!' clearly and angrily when a metal part broke in his hands. Usually he would say, 'My word!' The fact is that Ron truly detests profanity, and you could almost feel the resentment in him – especially if Valerie was present – when Peter Lake would roll off a casual four-letter fusillade. Once we were all in the mess hall talking, as we often did, about the cook's intolerable slobbishness, his bullying of the firemen and general vulgarity; we were trying to decide who could talk to Case with greatest effect short of a punch in the mouth. In a pause Ron said, 'Peter Lake can talk to him – they speak the same language.' It wasn't meant as a declaration of open war with Lake, whom Ron genuinely likes; it was just that Ron had finally found the spot to register effectively his distaste for Peter's language."

I asked Gimbel if he had gotten to know Ron during the expedition. The question made him uncomfortable, and he thought awhile before he answered it. "Ron's a very dedicated guy," he said, "and all he thinks about is his profession – for example, he only reads technical material, and can't understand why the rest of us bother with fiction." He paused again. "You know, I like Ron very much, I really do, but sometimes talking to him is like talking to that diving suit . . ." – here he pointed to the suit hanging on his cabin door – "there's nobody there." This sounded more critical of Ron than he intended, and he frowned, seeking to qualify it. "Of course, Ron got pretty excited during the marathon off Durban, but the only time he ever showed real emotion was the last day on the *Hermes*."

Despite a relatively successful night dive on the wreck – the night that the big rays were filmed – it was decided next day to abandon the *Hermes*. A wind had come up, the visibility had worsened, the current had resumed, and there was even speculation, based on the apparent absence of a flight deck, that this was not the *Hermes* after all. In the bad sea conditions, the job of letting go the lines proved long and arduous, and in the relentless heat, the low morale and general disappointment led to numerous flare-ups, despite Stan Waterman's attempts to smooth things over.

The anchor chain had been secured to a forward gun turret of the *Hermes* so that one blow of a hammer would cause the shackle pin to drop free. But there was also a valuable steel cable leading down from the *Terrier*'s stern, and this had jammed itself between the propeller shaft and the hull. Clarkson was rightly convinced that the cause was hopeless, but Taylor, who enjoys difficult problems, felt the cable might still be salvaged. He suggested a task for himself which involved swimming two lengths of the 450-foot hull; Clarkson, excited, exclaimed that no diver could do this without seriously overextending his time in deep water and risking the bends. Ron took this as a reflection on his professional experience. Staring out to sea, he said, "Maybe you shouldn't judge others by your own ability." In a cold way, Ron was furious, and Phil went stomping off, red in the face. A little later, when he answered a question from Ron with a jeering reference to Ron's self-confidence – "Why ask *me*? You're the one with the ability!" – Ron said coolly, "Phil, perhaps I should have said, 'I'm willing to give it a darned good try.'"

Ron tried and succeeded without mishap in the swim that Phil had said was not possible, and knocked free the shackle pin that released the forward anchor chain. But there was nothing to be done to free the cable, which was snapped apart at dusk by the *Terrier*'s engine; a hundred and fifty feet of the cable were lost. "All that hard and dangerous work for nothing," wrote Valerie crossly.

The person most endangered, as it happened, was an innocent bystander named Stuart Cody. Stuart had made his maiden dive only a few days before, when he had been mildly hit by nitrogen narcosis, a departure from reality induced by breathing nitrogen at high pressures. Cousteau likes to call it "rapture of the deeps" but actually it is rarely rapturous, and under its influence a diver may make fatal mistakes in judgment. Stuart was especially subject to narcosis because even for a beginner he gasped up an unusual amount of air, and on a subsequent dive he also had a mild attack. Nevertheless, on that last day, when he asked if he could accompany Ron Taylor on his long swim, Taylor and Waterman, who were in charge of the dive, did not dissuade him: probably they were too preoccupied to give Cody much thought and, in any case, Ron would much dislike the responsibility of refusing Stuart's wish. "I just take the orders," he often said; he thought of himself as a hired hand and tried not to exceed this concept of his role.

Having made the descent, Ron set off along the hull with Stuart in pursuit. Ron is a fast, efficient swimmer, and in his desperate struggle to keep up, Stuart was sucking up air at a dangerous rate. Very shortly he lost track of his situation. "I didn't feel turned on or anything," he said later, a little disappointed. "I was just depressed. I was wandering around down there, and everything seemed so gray and pointless that I couldn't understand why I was there. I just felt like giving up." Miraculously, it occurred to him to look at his tank pressure; his air was almost gone. He made his way to Waterman, who sent him up to Gimbel, who was safety diver. While Cody decompressed, Gimbel gave "buddy breaths" of his own tank until another could be brought down.

Gimbel recalls the whole day with chagrin. "We made a bad mistake," he says. "We were just too busy to keep an eye on Stuart, that's all, and a new diver had no business down there without somebody watching him every minute. As for Clarkson, you must understand that Phil was in a remarkably difficult spot. His job was not simply that of an ordinary production manager, whose work is basically organizational and administrative. On the *Hermes* Phil was also a first-line diver, working as hard as any of us all day and then making the trip back to shore each night where cables or letters – which were unjustified and were breaking his heart – would often be awaiting him from the home office telling him how rottenly he was performing. The fact is that he was handling two men's jobs, and what with the heat of Batticaloa, his marginal health, and the grueling trial the *Hermes* was putting us through, his nerves were shot and his mood was pessimistic. But Phil's problem is not one of too little gameness, but rather too much."

That evening the ship headed for Colombo, where it refueled and took on water for the 2500-mile voyage to Madagascar. "In Colombo, I had my lousy teeth fixed once again and splurged on jewelry. I shall glitter like a Christmas tree when I get home." Valerie thought Batticaloa picturesque, and was touched by the gentleness and courtesy of the fishermen who had come one by one in the days afterward to thank her for having shown a group of them her photographs of sharks. The Asian ways were a mild relief from the clamor aboard the *Terrier*, which is not to suggest that she didn't like the film crew; on the contrary, she once remarked that she liked all of them. "Usually there's *one* person that you can't bear, but not in this group. American men do strike me as childish, for all their good education, but I really like *everyone* on this expedition – that's rare, you know.

"And Peter and Stan are first-class divers – I'd dive with them anywhere. They know what they're doing every moment, they're *aware*; they won't get you in trouble by getting in trouble themselves." Ron agreed. "Peter and Stan are good, cool, efficient divers. Scuba, that is. I wouldn't know about skin diving."

The *Terrier* sailed for Madagascar after supper on June 20. The night was clear and the tropical sea was as silent as a mirror. Passing through the headlands at the mouth of Colombo harbor, Gimbel set off a batch of firecrackers out of sheer relief at leaving. Aboard ship the morale was high, although to compensate for the departure of Francis and Claus, the film crew was obliged to stand watches on the bridge. Valerie also did some of the cooking, since Case, with only the crippled Lindsay to help him, considered himself overworked, and allowed his cuisine to fall below the low standard he had set for it. Certain meals were so bad that "even Gimbel wouldn't eat them" – Gimbel's indifference to the food before him was a standing joke.

The only person who did not stand watch was Stuart Cody, who already had too much to do and whose habits, in any case, did not lend themselves to schedules. Cody's resistance to posted rising hours and mealtimes was decried at first by punctual people like Clarkson, whose business it was to maintain some sort of efficiency; on the other hand, he would often work until after midnight, and he was so valuable that in the end he was encouraged to follow the private schedule that suited him best.

Cody, who has a successful electronics firm in Cambridge, Massachusetts, had not come along on this expedition for adventure or escape but because the variety of equipment in his care would be a challenge.

*Above*: Whale-catcher W–29

*Below*: W–17 with dead whales alongside

Arvid Nordengen, Captain/Gunner of W–17, who harpooned "an enormous shark" two or three years before the expedition

The divers and the cage: *above from left to right*:
Peter Gimbel, Valerie Taylor, Ron Taylor and Stan Waterman

*Above:* Peter Gimbel and white-tip oceanic shark

*Right:* White-tip reef shark

*Below:* Valerie Taylor baiting at Europa in the Mozambique Channel

*Left:* Jim Lipscomb, co-director surface photography

*Below:* Stuart Cody, sound

*Left:* Peter Lake, still photography

*Below:* Tom Chapin, grip

Peter Gimbel, producer and cameraman

Stan Waterman, associate producer and cameraman

Peter Matthiessen, chronicler and spare hand

*This page and facing page*:
Filming the great White shark

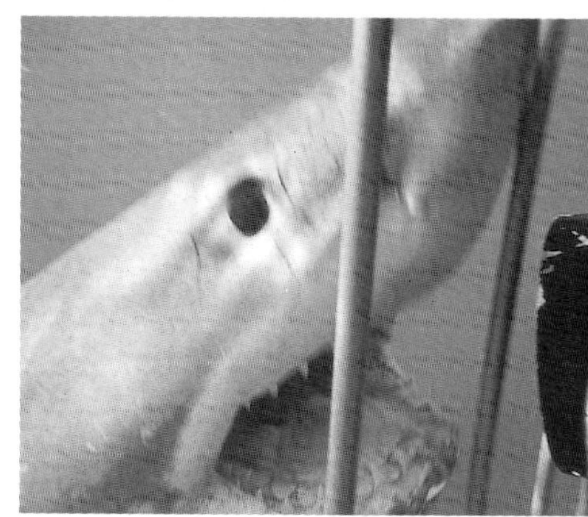

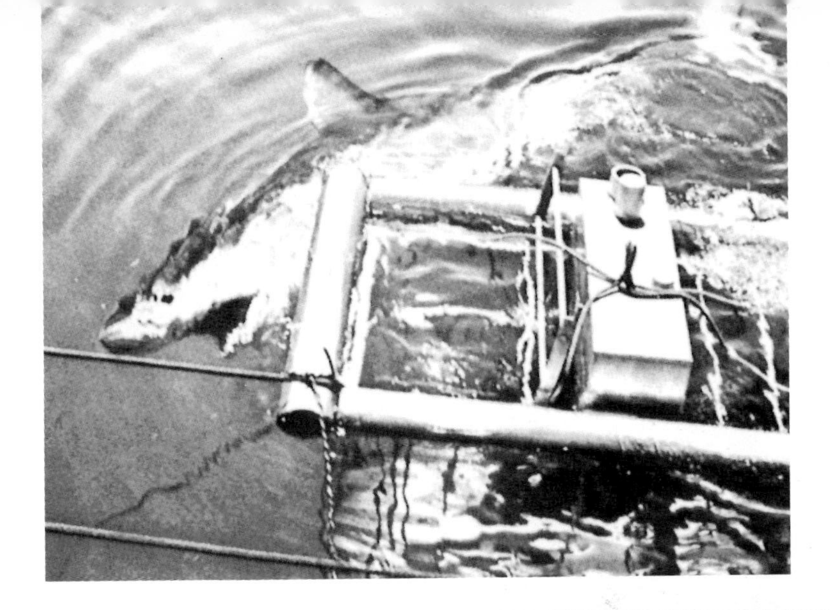

Peter Gimbel filming in the cage

The challenge turned out to be so constant that Tom Chapin had to take over as sound man for Jim Lipscomb's camera so that Stuart could devote full attention to more complex jobs, including the electrical system of the entire ship. The *Terrier* had been given a minimum of maintenance and was falling apart, and gradually, as her chief engineer lost interest in his ship and went into retirement in his cabin, it devolved upon Cody to keep the old hulk going. Both crew and officers of the *Terrier* had hated Ceylon, and by the time she sailed out of Colombo the ship was filthy with neglect. Most of the officers and firemen were drinking hard (the deck crew, on the other hand, was entirely able and dependable), and every man, without exception, longed for home. Among other things, they were short-handed and overworked, and their situation worsened when the first calm days of the southward voyage were blown away by 25- to 35-knot gales of the southeast monsoon, which dogged them all the way to Madagascar. One night Gimbel, clutching his dog, went right over a rail during a 30-degree roll and fell from the bridge ladder to the pilot deck below; he hurt his hip, elbow and shoulder, but he could as easily have gone over the side.

June 23. "Horrible, horrible weather . . . I was up on watch in the wind and rain when Sam, the one-eyed crewman, came up and said, 'You shouldn't be out in such weather, Mrs Taylor, let me take over.' He is such a sweet man. I have become very fond of Sam. He may be coloured but to me he is pure white. He was a wonderful help during the *Hermes* jaunt, helping me onto the boat with my twin tanks weighing me down and making me helpless. He is so strong. He can pluck me from the water, tanks and all, with just one hand."

Sam Lloyd was a stumpy man who dressed invariably in street hat, shorts, and the long knee socks of the white colonial; his solitary eye was set in an ageless head that might have been hacked out of old mahogany. The divers were grateful for that sleepless eye, which was always on watch from some part of the ship, and had picked out bobbing heads over and over.

June 24. "Made a trifle for tea. The Americans loved it . . . Sometimes I wonder what I am doing out here, what I am trying to prove. My mother brought me up to get married and have babies but something went wrong and to her I have been a disappointment. I married, but only to follow my husband through one adventure after another, never settling for long . . . Can't go on forever like this – will be 34 next November. Surely there is more meaning and substance to life than this . . . Oh, well, maybe next year I'll do something different . . ."

June 25. "The weather is a fair cow, really. The swells are getting bigger each hour . . . Everything is sliding backwards and forwards. I have blisters on my hands from the wheel . . ."

June 26. "Won't these rotten seas ever calm down, won't they ever still? I feel I have been rolling all my life . . . Spent eight hours on watch today – very boring indeed. Sam said he didn't think Francis would be made white. God, I hope he is wrong. Poor Francis, how I feel for him. A better man than most, damned because of the colour of his skin. The whole system stinks stinks stinks . . ."

June 27. "Another day. We are still rolling. I wonder whether Francis and his wife and children woke up coloured or white. I don't believe in God, but I prayed for him in case he does, or is God really only for the white? Maybe he is really the devil and has us all fooled."

June 28. "Bad weather still with us . . . We hope to pick up Francis again in Madagascar. For a man to spend his life savings to fly from Colombo to Cape Town just to be told he isn't white and never can be is too terrible to think about . . ."

That night the *Terrier*, fighting gusts up to 50 knots, made port in Madagascar, passing rapidly from the rough seas through the narrow heads into the smooth bay of Diégo-Suarez.

It was winter in Diégo-Suarez, which lies 12 degrees south of the equator. This pretty French colonial town on a ridge above the bay lacks the humidity of Ceylon and can claim a good French restaurant. Lipscomb, Waterman and Chapin found a tennis court and resumed matches begun in Durban, while the Taylors and Peter Lake, off on an inspection of the town, were soon befriended by two local ladies who paid cheerfully for their drinks and taxis, since the visitors lacked Malagasy francs. "They kept petting me, and relieved themselves on the ground right in front of everyone. I was most surprised and so was Ron. But I liked them, for they were very gay and good-natured."

* * *

While much better, Phil Clarkson was still not up to sea travel and poor food, and he went to Madagascar by air. On June 30 I met him in Nairobi, where he and his family had stopped over for a few days, and we flew that morning to Tananarive in Madagascar, continuing north to Diégo-Suarez the following day. Phil told me the details about the *Hermes* debacle that Peter's letter had left out, and by the time we rejoined the others, at midday on the first of July, I expected to find an atmosphere of gloom.

The film crew was at noon mess in the galley – the ship was to sail in the late afternoon – and to judge from appearances their spirits were very good indeed. Credit for this must go to Gimbel, who is one of the best expedition leaders I have ever worked with; many are intelligent and conscientious and hard-working, but few are so courteous and considerate to their people, especially when things are going badly.

In private, however, Peter's stress was evident. "I wish to God I'd never heard of that bloody ship," he said passionately in his cabin, and in the days that followed, the rest of the film crew said much the same. Jan Moen merely smiled ruefully, shaking his head at the mention of the *Hermes*, yet unwilling to venture an uncharitable opinion even with Ceylon twenty-five hundred miles behind him. But Anson Lloyd, a deck-hand who had been ordained a white just before leaving South Africa and was rewarded for this new eminence by being promoted to second mate, spoke for the whole ship's company in his condemnation of Ceylon; as a new white man, he insisted with some vehemence on the low character of the Ceylonese "natives".

It was plain to see that the ship's condition and morale had not improved, and there was a decided sag in discipline. The voice of Jan Moen, never a forceful one, had been muted by the presence of the vessel's official master, Knut Paulsen, even though Captain Knut, like the chief engineer, emerged rarely from his cabin. With the *Terrier* short-handed, men came and went more or less as they pleased, knowing that they would not be fired with the ship so far from home. Only one of the two new firemen, taken on when the ship left Durban, was of any use whatsoever, and it is a sign of the state of affairs aboard the *Terrier* that the other, an implacable alcoholic, was the man sent ashore in search of his shipmate Philander, who had set out just before sailing time in quest of drink.

Because Diégo-Suarez is the chief bunkering port in this part of the Indian Ocean, it is ridden with low sailors' dives, and as far as could be told by Gimbel and myself, who followed their labyrinthine path all evening, Philander and his friend did not miss one of them. The ship's sailing time was delayed five hours, with the result that she did not arrive off Astove Island at daybreak, as had been planned, but just before noon. Four working hours had been lost on a location with production costs of five hundred dollars per working hour.

On this leg of the voyage I was assigned to the cabin of Stan Waterman, for which I was grateful in many ways. The cabin was the most pleasant on the ship, not because of its location (it was just aft of the smokestack) but because Stan kept it fiercely clean; warmed by his pipe rack and his pictures of his pretty family and a well-worn copy of *The Wind in the Willows*, which he knows by heart, it was an oasis of gracious living in the sooty squalor of the *Terrier*. Once he had recovered from the shock of my intrusion and made certain that I understood his rules of housekeeping, my host was hospitable to a fault, and we greatly enjoyed our evening rum, regaling each other with tales of bizarre experiences in what Stan delights in calling "the exotic ports of the Levant".

After graduating from college, Waterman entered the State Department, but though diplomacy is second nature to this man who dreads discord of any kind, he shortly abandoned affairs of state in favor of a blueberry farm on the coast of Maine. For four years he was a farmer on Penobscot Bay and later on Prince Edward Island in New Brunswick, and during this time became interested in diving, taking on odd diving jobs up and down the coast. In 1954, the same year that I went into the charter boat business at Montauk Point, he built a charter boat and for the next four winters, from Christmas until June, took fishing and diving parties out of Nassau. In these years he met Peter Gimbel, whose dive to the *Andrea Doria* he had much admired.

Stan's amateur films of the Bahamas reefs were soon in such demand that his free shows became paid lectures; he sold the boat, and has been an underwater film-maker and lecturer ever since. *Genesis*, an excellent short film on underwater life, was condensed from *The Running Tide*, which he was shooting in Bora Bora, in 1965–66, when he had his near-fatal attack of the bends. "I was just careless about my time," he says, laughing at himself; instead of dwelling on his close escape, as most men would, he prefers to speak of the many kindnesses he received during his experience and the good wine drunk in celebration of his emergence from the recompression chamber. Stan doesn't care for introspection; he has hit upon a life style that contents him and gives pleasure to everybody else, and that is enough.

Astove Island, a hundred and forty miles northwest of Diégo-Suarez, is a southern outpost of the Seychelles archipelago; it lies five hundred miles southwest of the island capital at Mahé. Astove is a true atoll, an ellipse of coral three and a half miles long by two and a half wide, surrounding a marly lagoon. This lagoon was once inhabited by dugong, and the high ground is occupied intermittently by man.

The man there at present is Mark Veevers-Carter, who has with him a wife and two children, as well as a small community of Seychellois laborers. Veevers-Carter, whose mutton-chop whiskers give him the aspect of an old sea dog, is restoring the run-down copra plantation on Astove, and has plans – problematical but fascinating – for diking and draining the lagoon and constructing a big rice paddy with the help of the plenteous fresh water of the wet seasons. (The sad word has just come that Mark, ill with infection spreading from bad teeth, was taken to Mombasa by a passing freighter and there died.) On the afternoon of the *Terrier*'s arrival he and his wife, Wendy, took us on a tour of the new house being built of island limestone, and the next day they invited us to lunch on the porch of the old one, an airy palm-thatched frame dwelling of the sort found on tropic islands throughout the southern seas. The family has ducks, chickens, pigs, cattle and a rabbit, all of which live beneath the trees between the veranda and the sparkling tropic sea; fish, shellfish and crabs eke out the stores of canned goods brought periodically by island traders and visiting vessels. From the garden come domestic sweet potatoes, tomatoes, coconuts, Chinese cabbage, radishes, parsley and green onions. A turtle grass, *Caulerpa*, which is cast up on the beach in great amounts, is gathered into compost heaps and makes a very good manure after three years of being leached of salt by rain.

At lunch, a fine curry was accompanied by barracuda paste and chicken, and we ate sea cress from the lagoon shores, wild purslane and a salad of wild cucumbers and *brèdes*, or spinaches, several species of which are indigenous to the island. Rum and lime juice were available, or iced tea – the house has a generator and a freezer – or palm wine. Fresh palm juice is taken in the morning from half-coconut cups hung under the inflorescences, which are first bruised in a way that makes them drip; the pint or more drawn from each tree ferments by noon.

Exploring Astove, I was stricken with longing for the solitude and peace of such an island, where I would live in a manner still more simple and self-sufficient than the one chosen by the Veevers-Carters. I discussed such a life with Valerie, who said she was content with her house and garden in the Sydney suburbs. All that was lacking in her life was children – "I think children would be nice for Ron," said Valerie, who had suffered a miscarriage in her brief first marriage. It would be an adventure, of course, to live on a desert island, but she thought the time for this had passed; at thirty-three she felt too old. She wasn't, of course, but I realized that I was. With a forty-year accumulation of responsibilities, I would do it now for the wrong reason. For me, the life-long dream of my own island would no longer be adventure but escape.

That afternoon, and for the next two days, we explored the reef. On the northwest shore of Astove, protected by land from the southeast monsoon, there emerges at low tide a bench of warm coral pools stalked by the reef heron. Two hundred yards out, the coral wall drops straight down a quarter mile, so abruptly that the bow of the dinghy nuzzles the coral heads while the stern hangs over a jet-blue that is almost black. The rim of this underwater cliff is split by deep clefts that lead back toward the beach, and under the lip are ledges and caves swirling with fish. Offshore swam mating pairs of the green turtle, and on a reconnoiter that first afternoon I saw a big male, three hundred pounds or better, gliding down the sea cliff in the crystalline light like a great bird in a twilight canyon.

Stan Waterman, who has dived more reefs than anyone aboard, thinks that Astove is the most spectacular he has ever seen. In his experience, only the Tongue of the Ocean, in the Bahamas, falls away so steeply into the abyss, and it cannot compare with Astove in the abundance of marine life, nor in the sunlit clefts and soft-shadowed caves inset with golden crusts of coral, glistening like geodes in the sun of the equator – a striking setting indeed for the large sharks that are said to abound here. According to Veevers-Carter, Astove's waters are alive with giant hammerheads, of which the Seychellois are deathly afraid; since none of these monsters had been seen, it was decided to try attracting them. The best-known shark attractors are speared fish, which send vibrations as well as blood into the current; we also tested a low-frequency sound projector (sound carries farther and faster in water than in air) which imitates injured fish mechanically and has had good results in bringing sharks. It turned out later that the sound projector was defective; whether or not this was a factor, no sharks came.

While waiting, I wandered down along the cliff face, staying close to the shelter of clefts and caves. The sinking sun illumined the underwater wall, and I drifted languidly in and out, attended by magnificent arrays of demoiselles and butterflies, parrotfish, sweetlips, tangs, grunts, unicorns and groupers. There was a squirrelfish with white points on its dorsal spines, and the big green Napoleon wrasse with its bulbous head, and Moorish idols, and startling triggerfish with big white egg spots on a black belly.

Overhead, engaged in spearing fish, the Taylors had abandoned their scuba tanks and were demonstrating for the cameras the skin-diving form that has made them champions. To watch them from below was like watching a sinister dream. First their surface silhouettes came out over the rim, long-legged, fin-footed, like trim black frogs outlined on the sparkling water. Then down they came, spear guns probing out ahead of them like antennae, down, down, down with slow neat fin flicks to thirty, forty, fifty feet. There they remained, exploring the caves for seconds, a minute, on that single breath, before rolling gracefully out and upward, never hurrying, ascending like black angels to the light. If one of the scuba divers was nearby, a breath might be taken from his regulator, and the hunt continued. Sliding forward with minimal motion, Taylor is deadly; one flinches at the tin ring of his spear gun fired underwater. Impaled, the fish shivers on the spear, turning and twisting as the string is drawn in, to die finally in quick mortal flutterings like a bird. As Waterman says, "Ron *stalks* those fish – it's beautiful. He's like a predator." Ron had discovered a truly enormous moray eel – by his own conservative estimate six feet long – and he lured this thick-bodied brute half out of its crevasse and fed shreds of giant *Tridacna* clam into its gasping jaws with his bare hands.

In the three days at Astove, there was a running dispute about whether or not to drift the cages down the reef. Some felt that using the cages would be superfluous and fake, since white sharks are not likely to occur here, while others mentioned the big hammerheads, and the imminence of mystery that the cages would lend to the dark undersea cliffs. The point was academic, in any case, since Jan Moen and Captain Knut, accustomed to the cold rocky coasts of temperate latitudes, would not believe that the bottom is a quarter mile down only a hundred yards from shore, much less that it could possibly be safe – though larger ships have done it – to lodge an anchor on the reef edge and let the unrelenting wind of the monsoon hold the ship offshore. The ship stood by a mile at sea, where it was much too deep to anchor, circling all night long in the same spot.

Day in, day out, the monsoon blew at 20 to 25 knots, moderating slightly in late afternoon. The filming went badly because the reef had a northwest exposure and therefore lacked good light before midday; at the same time, a flood tide was required, since on the falling tide the water was roiled and murky with the algal bits, fine sand and sea waste that poured off the coral shelf. Light and tide never came in the proper combination, and meanwhile the big moray moved to parts unknown, the mating turtles scattered, the giant groupers became wary, moving offshore when approached, and a bottle-nosed whale (a true whale, *Hyperoodon*, not the bottle-nosed dolphin) that played about the ship at noon on the 4th of July came and went unrecorded by the cameras. So far not one shark had appeared, and it was decided that the *Terrier* would sail that evening for Aldabra. In the afternoon an attempt was made to salvage some continuity footage from the reef, but the wind was increasing, and flying clouds spoiled the light in shot after shot.

That last afternoon I went exploring, coasting down below one hundred feet; here the last color, a brown-fringed overlapping coral like shelf fungi, died away. Beneath was bare dark dungeon-colored wall, the bastion of old reefs of other ages, and when, far above, clouds passed across the sun, this nether world was plunged into cold gloom. Then I would peer up the sheer cliff, searching in vain for other divers. Seeing none, I turned outward to the void, expecting the great hammerhead to rise out of the deep. There were no sheltering caves, no clefts, only the gray wall and the blue mist, and instinctively I wriggled upward toward live coral. Then the clouds would pass, and I sank through rays of light like a particle in eternity, hearing the ring and tock of unknown voices and the closing and opening of my own heart. When apprehension eased, I would probe down again into the abyss. I did not know what I was searching for; nothing was there.

Aldabra lies at latitude 9° 24′ south and longitude 46° 20′ east, two hundred and sixty miles northwest of Madagascar and four hundred miles east of Africa. It is composed of fossil reefs, uplifted from the sea perhaps a hundred thousand years ago, and its north side is a wall of limestone ironshore, broken by rare small beaches. Aldabra is an atoll, one of the world's largest, and since its rim is nearly complete, it resembles a true island. At dawn, dark frigate birds were already aloft on the hard rush of the southeast trades that sweep across the island from May through October, and terns dipped in the blowing spray off the mile-wide mouth of Grande Passe, the main entrance to the atoll's great lagoon.

The *Terrier* came up under the west end, where a sheltered beach is shaded by wind-leaned palms. Here a research station has been established for scientists interested in the island's unique fauna; even in Darwin's time, Aldabra was the last refuge of the Indian Ocean giant tortoise, and it also claims such relict creatures as the flightless rail.

A young British biologist at the station had bad news. Like Gimbel, he had come here because he had read of Aldabra's shark hordes in a book called *Beyond the Reefs*, by William Travers. Ecologically, he reported, Mr Travers' book was full of holes; he had seen one modest shark in the past three months. Even the manta rays were few, and the turtle beach off which great herds of mating turtles had been said to gather was the one beneath his feet; with the coming of the scientists, the turtles had gone elsewhere.

Gimbel looked stricken. Though we had just gotten to Aldabra, he announced that next day the *Terrier* would head back to Diégo-Suarez to take on fuel and water, then proceed straight to the Comoros and St Lazarus Bank. Big tigers and hammerheads were said to be plentiful at various locations in the Mozambique Channel, and probably whale sharks, sea turtle, manta rays and dugong.

Peter's logic was inevitable, but most of us were stunned; the 500-mile round trip from Diégo-Suarez to the Seychelles would be written off as a near-total loss, and the prospects for the weeks to follow were very vague indeed. No one had information, much less hope, that the mountainous Comoros could produce the large marine creatures that the film needed (the coelacanth for which the Comoros are famous is a deep-water creature that has never been seen at depths attainable by divers), and the St Lazarus Bank, a submerged sea mount far off northern Mozambique, would be rough, roiled and unmanageable unless the trade winds moderated. If the monsoon patterns extended south into the Mozambique Channel, the water clarity on that windward coast would also be poor, and after so many disappointments – no first-class footage had been obtained since May 5 – it could no longer be assumed that the promised sea creatures would be there.

That afternoon we went ashore at the East Pass to film the giant tortoises. In this region of Aldabra, the tortoises occupy a high limestone ridge that forms a rampart of the northern coast, and their grazing of the native grasses has created a lovely woodland pasture shaded by casuarina and set about with a tropical shrub of acanth and loosestrife: the pasture overlooks the shallow reef ledge and the sea. Overhead, seen through airy casuarina, the frigate birds from the rookeries in the lagoon cross the bright sky.

As the ship's tame naturalist, I was to play opposite Valerie in her Adventure on a Desert Island, marveling at flightless rails, thumping tortoises, and wandering breathless through the magic wood. But that afternoon was the culmination of a running dispute I'd had with Lipscomb on the subject of dark glasses, at the end of which my chance for immortality as a movie star was gone forever. Lipscomb feels strongly that people in shades come out on film with a total lack of personality, and therefore, in the fierce glare of tropic oceans, the principals of this movie must squint like mummies instead. But I was not a principal, nor even a full-time member of the crew, and my dark glasses have a prescription in them. Besides, I was in ill humor at the prospect of leaving Aldabra so soon after arrival and at having to see my first giant tortoises and flightless rails in the disrupting presence of a film crew, and I took out my irritation on the whole Adventures-of-Valerie approach in general; it seemed to me that the surface footage depended too heavily on Valerie's lovely face and professional acting ability in a film where acting had no place, and that these winsome excursions all over the shores of the Indian Ocean would slacken the tension of the film, and rot it with cuteness into the bargain. In addition, I was irritated by my own secret wish to appear in the movie, so that when Jim yelled at me to take off my shades, I snapped crossly that I had waited for years to see the creatures of Aldabra, and didn't want to be in his movie badly enough to reel around the place half blind. Finally we compromised. When he yelled, "Glasses!" I would hide them in my sombrero – another bone of contention – and drift past his lens like a stuffed shark. (In the end, the tortoise footage came out dark as well as stilted, and was discarded.)

On an ocean expedition there are few chances to be alone, and I seized the first opportunity to wander off by myself. From the dancing shade the tortoises watched me – I saw perhaps a dozen altogether – the light glinting on the ancient gray metallic heads. Although these animals have long since disappeared from all the continents and no connecting population occurs anywhere between, they are so closely allied to those halfway around the world on the Galapagos that both populations were formerly assigned to the same genus (*Testudo*).

At midday the tortoises lay like boulders in the filtered shade, but later they moved out to graze, arching small heads on long necks to eat the grass, and bumping over the roots and limestone with an old hollow ringing, like footfalls on the far side of a cathedral. These contemporaries of the dodo have fat round feet like those of elephants, and bronze scales with burnished rings, like the age rings in a cross section of polished wood, and their fossil bones are almost as old as the island of Aldabra.

White-flowered caper grew, and amaranth and nightshade and verbena, and from the windy trees came the sweet ringing of a rail. The flightless rails resemble small, neat pullets, and their rich chestnut feathering, washed with green, is iridescent in the wind-bared light of ocean afternoon. The birds are tame, entirely trusting, as vulnerable to the sticks of men as the great slow tortoise. Under the clear gaze of such creatures, in this bright whispering wood, there comes a painful memory of Eden.

NEXT MORNING THE film crew's nerves were tense, and growing tenser. Diégo-Suarez was thirty-six hours away, but in order to get there at dawn on July 8, it was necessary to sail at noon, since the ship had to stop at Astove for a crucial zoom-lens battery that Jim Lipscomb had left behind. Gimbel wanted the divers to drift underwater through the swift tide channel of East Pass, which abounded with fish and might produce a shark, but first the shore crew would take advantage of an early tide to enter the shallow lagoon and film the rookeries. Both small boats were required in the lagoon so that Lipscomb, in one, could record the explorations of the other. Meanwhile, the divers were stranded.

The Aldabra lagoon is a pale waste of white tide flats and clear mangrove creeks where nests of frigate birds and red-footed boobies overhang the water on all sides: in the mangrove green, the red throat pouches of the male frigate birds gleamed like apples. Fairy terns and the blue pigeon raced overhead on the sunny wind, and white-eyes and drongos came confidingly to the low branches.

This tranquil scene was marred almost immediately by *Homo sapiens*. Cody, straining to record the wind-thinned nackering of frigate birds, became furious at Lipscomb, who broke into the sound track with a question about frigate-bird plumages that could have been asked later. In the course of a loud argument Lipscomb told him to grow up, whereupon Cody dropped his earphones and switched boats. Tom Chapin, who usually played straight man in Valerie's escapades ashore, took over on the sound machine, and Lipscomb remained in the lagoon long enough to shoot a second magazine. By the time the boats got back to East Pass the tide was already slackening, and the divers, who had been stranded aboard the *Terrier*, arrived at the channel too late to film the tide-run sequence that Gimbel wanted.

Peter was incensed, and could not hide it. "Jim, yesterday you said I was getting desperate, and I am! You know how badly we need the underwater stuff, and you kept those boats in there, and now we have to get back to Diégo, and lose still more time stopping off at Astove for your bloody battery!"

Lipscomb received this calmly, without comment, merely frowning a little as if Peter had spoiled his concentration on something else; he is an intent man, a professional, and a survivor. Jim is affable and handsome, and a stalwart in the boisterous camaraderie of the mess room, where he is often called "Big Jim." But in the hearts tournament that takes place almost every night, the shaking of his hands when he holds good cards suggests the intensity behind his easy manner. One of his friends feels that Jim does not trust people; his answer to any question is apt to be preceded by a brief stare, as if he were measuring a risk.

Gimbel, bitterly frustrated, could barely be persuaded to take a quick look at the tortoises. Afterward the entire film crew, with full diving and camera equipment, was squashed into the two small boats for the return journey. Outside the pass where the outgoing tide collided with the wind, there was a heavy rip, and the *Terrier*, predictably, was not only well eastward of the channel but two miles offshore. We beat upwind to her.

The sea was rough, and coming alongside, the boats banged up and down on the steel hull; at the same time a hot brown stream of water from the bilges sprayed the film-makers, bringing artistic tempers to a boil. Clarkson tried to bring some sort of order to the unloading, but his nerves, which tend to respond to Gimbel's, had been tightening all morning and he cursed wildly at Peter Lake for moving too slowly. Though Lake handled it well, it was an ugly episode that depressed Gimbel, who was already worried about Phil's health and deteriorating spirits.

Within minutes Clarkson had apologized, and at the mess table shortly afterward, almost everyone was in good spirits once again. The exception was Cody, who expressed more resentment of Lipscomb and was answered in no uncertain terms. Before Stuart could respond, he was cut off by Stan Waterman, who put on his comic mask of goofy evil, saying in a stage whisper behind his hand, "Jim really laid *into* you, Stu – you're not going to let him get away with *that*, are you?" Cody stifled a shrill retort and managed a laugh.

Waterman knows just how and when to play the fool, saving awkward situations by drawing attention to himself, and his diplomacy has averted ruptures over and over. Skillfully he turned the conversation to the matter of the great green Maori wrasse, shot yesterday, that now lay sun-charred and stinking on the deck: Lipscomb required a continuity sequence of a diver cutting up fish for chum, and Ron, Stan and Peter had an odds-or-evens contest to see who would get this unpleasant task. The loser was Peter, who laughed in a queer quizzical way into his plate when Stan professed to have rigged the loss by guessing Peter's psychology in advance. The laugh didn't sound right. Gimbel caught my glance, and we both looked away.

The chum sequence was shot right after lunch, and by 3 p.m. the ship was under way. Soon Cape Hodoul, named for a buccaneer who plied these coasts, fell away on her starboard quarter. I was out on the stern watching Aldabra sink into the sea when Phil turned up and sat down heavily in a tattered deck chair. He looked awful. His ulcer was acting up again, the food was greasy and the seas were rough, and he was just barely hanging on. We discussed the morning – he and I had also had two minor skirmishes – and he commented, without complaining, that it was hard to stay good-tempered when he felt so sick. Possibly he would fly ahead to Lourenço Marques to avoid the sea voyage across the Mozambique Channel. I said that if he didn't he would wreck himself, which would not do the expedition any good. Last summer Phil lost both parents in a horrifying tragedy, and since then has not given himself time to recover his balance; his hard and thankless job had been made much harder by his own lack of resilience and his bitter refusal to tolerate his own frailty. "One likes to think one is tough enough to take anything," Phil said sadly, and I nodded; there was nothing to say.

As the afternoon wore on, the wind increased, and by supper half the film crew felt sick. Astove lies eighty miles east of Aldabra, and we had hoped to reach there that evening before the Veevers-Carters put out the light that we needed as a landfall, but the ship was beating straight into the monsoon wind, the night was overcast, and Astove is low and small.

During the night of July 6 the seas grew so high that the *Terrier* was forced to cut her engine to half speed, or six knots. Two huge swells stopped her dead in her course with an impact that drove her occupants into the heads of their bunks. At dawn Jan Moen reported that we lay off the Cosmoledo atolls, still eighteen miles west of Astove, and already a half day late at Diégo-Suarez. A four-knot current and thirty-knot wind had held the ship to a ground-speed of less than five knots. In the rough seas, Lindsay Clarkson had fallen and reopened his raw hand, and was seasick as well; he was lying in his bunk, doped up with painkiller.

At breakfast Valerie said in a quiet voice, "It looks like the rest of this trip is going to be a trial." She was discouraged by the repeated failures, exhausted by the long rough night, and wearied in advance by the long day and night ahead. Fortunately Peter wasn't present when she spoke, and certainly she wouldn't have said it if he had been, because the expedition had no choice but to continue and hope for the best.

The undermanned crew was as exhausted as the passengers, and Stan and I volunteered to stand the morning watch. The bridge rolled and the laboring hull sent out hissing sheets of white, but sea and sky were a hard, wild blue, with sooty terns dancing everywhere over the white-caps, and a lone booby rifling downwind; toward noon a tropic bird, bone-white, circled the rigging like a firebird from the meridional sun that lighted its wing tips and the long tail feathers streaming out behind.

The horizon on all sides was empty. Astove should have been sighted by 8.30; we had somehow missed it. For an hour Jan Moen kept appearing on the bridge, scanning the horizon with binoculars and shaking his head at the queer ways of the sea. Prodded, he acknowledged that he did not know where we were; perhaps what he had taken for Cosmoledo had been Astove. Soon Gimbel was standing there in a pale windbreaker looking strangely shrunk, an invalid, in dark glasses and soft hat. He greeted with dead silence the news that the ship's position would be uncertain until a sighting could be taken at noon; no one had to explain that meanwhile we might be steering farther and farther off our course. Now Captain Knut, blinking uneasily in the bright light, left his cabin to join Jan in the head-shaking; they persuaded each other that Cosmoledo had certainly been Astove. Due to the set of the current, the ship had passed well to the south of Astove, which on this ragged wind horizon was too low to be made out by eye or radar.

At noon it was discovered after much computation that the ship lay thirty miles southeast of the island, toward Diégo-Suarez; it was a five-hour round trip back. "That was a very expensive mistake," Gimbel said finally, stalking out of the chart room; it was hard to tell whether he was speaking to the ship's officers or to Lipscomb, whose forgotten battery was the purpose of the trip. Jim, standing in the doorway, made no answer. But at lunch Peter seemed cheerful enough, making the best of it; he laughed as hard as anybody when Cody, hearing that a cache of strange phials had been discovered in a bulkhead cranny near the cabin of the chief engineer, cried out, delighted, "You've discovered the Chief's chemistry set!"

Peter had indicated a desire to talk, and in his cabin that afternoon he let off the steam that had been building in the past few days, including a simmering resentment of those members of his crew who had been unwilling or inept or were otherwise jeopardizing his two years of careful preparation. But all of this was said in the context of what he had written to me earlier, and still believed, that "there is no member of this crew I would exchange for anybody else." It did him good to bitch a little, but he was overcome once more by his fear of failure, and was suffering the inevitable attrition of self-doubt. At one point he asked if there was any widespread gripe or disaffection in his crew that he should know about, and I said that he had the complete respect and loyalty of everyone, without exception, and that rather than doubt it, he should feel free to draw on it by delegating more responsibility, especially now that Phil Clarkson was leaving the ship.

"Yeah," he said, nodding. "Yeah." Most people get depressed when they let their guard down, but Peter seemed relieved and even laughed a little. "You know something? You know how paranoid I was getting? Well, yesterday, when we did odds or evens to see who had to cut up that goddamned smelly fish and I lost, and Stan said they'd rigged the throw against me, I almost believed him. I wondered all afternoon if something funny was going on; I wondered if those guys were out to get me." He gazed at me, shaking his head. "Let's have a drink," he said. "I feel much better."

Yesterday Peter Lake described how he had met Gimbel in 1963 at the parachuting grounds at Orange, Massachusetts; at that time Lake was still an undergraduate at Dartmouth. Subsequently he accompanied Gimbel to Peru, where Gimbel had been much more relaxed, he said, and very much in charge of the situation; here he was on less familiar ground, with a good deal more at stake. In Lake's opinion, Gimbel is simply too gentle and decent to direct an expedition that is also a complex commercial operation; furthermore, he does not know how to delegate authority, so that too much of the nagging details had devolved unnecessarily upon himself.

Lake has more insight into Gimbel than anyone in the film crew, and his observations are worth repeating:

"Gimbel has so much at stake here that failure would be tantamount to a spiritual demise. In spite of the extraordinary achievements of the man, he has yet to 'become something,' a feeling not requisite perhaps to members of my generation, content to be doing their thing, but a belief I feel deeply ingrained in Gimbel to push him toward 'becoming something.' He has, of course, been variously shoe salesman, management trainee, investment banker, and underwater something-or-other, plus the 'adventurer' (what an odious word) that I met him as. Rather, he has participated in those activities, but has not *been* any of those things. I am not entirely convinced he is a film-maker, either, at least not in the modern sense of being able to manipulate the plastic qualities of film. Rather, he uses film for self-expression, an older and certainly more respectable use of the medium. He is not so enmeshed in film, however, that he could not use another medium were it better suited to his ends. He is trying to make a personal statement and build a movie around his own experience . . .

"The strain shows clearly, especially when the movie goes badly, as it did in Ceylon. We are so much the victims of nature that at times our situation seems hopeless. We depend on water clarity, wind, waves, light, marine life, complicated equipment, a half-assed ship owned by a pirate, and many people whose health is continually under great jeopardy. To have every element working at once is a luxury we seldom enjoy. In Peru all we had to do was keep healthy and keep moving; everything else would take care of itself if we did that. The adversities – cold, insects, hunger, fatigue, wetness, homesickness – were shared to nearly the same degree by all. Here, all have different problems, and ultimately they fall back on Gimbel. Ron and Valerie are the only two members of the crew who have not, as far as I can tell, put any burden on Peter. Some have done so in ways they may not recognize, but they most assuredly have, although to a greatly varying degree. Thus he is burdened not only by his own considerable worries, but also by those of the rest of the production crew and of the ship . . ."

In the afternoon the ship's morale improved. The *Terrier* churned downwind to Astove and got away from there quickly after retrieving the battery; meanwhile, the wind had moderated so that the ship was able to maintain speed on the eastward bearing. Also, word had been received by radio that Francis deNikker, the crewman who had returned to South Africa for his reclassification test, was awaiting the ship in Diégo-Suarez. And the chief engineer had turned up again, his old shy kindly smiling self, after a long bout with his demons in his cabin.

At dawn on July 8, Madagascar lay off the starboard bow, and by 8 a.m. the ship had rounded Cape d'Ambre on a calm ocean and was coasting south down the lonely tip of the great island. Toward 10.00, she passed through the headlands into the beautiful open bay of Diégo-Suarez, once a haven of pirates. Soon the pilot's launch was alongside, and the *Terrier* tied up finally at 11 a.m., four hours behind schedule.

Francis was waiting on the dock with the new messboy, Clive, who looked just like Claus grown two inches taller, and was just as easy and good-natured. Answering a barrage of questions, Francis said he had had to leave Cape Town before the final judgment was handed down. Though he was smiling and expressed great hope, he was not yet white. (Francis is at present a crewman on W-17, and a letter from Captain Nordengen dated August 13, 1970, more than a year after Francis' appearance before the Appeal Board, informs me that he is still awaiting the State's judgment on his fate.)

The day was spent in furious organization so that the ship might sail that night. Ship's agents, chandlers, taxi drivers and a party of whores came swarming aboard, and the girls remained for the best part of the day, plying their trade in the iron rooms belowdecks. The film crew had an excellent French lunch at the Hôtel de la Poste, where two chameleons were caught in the branches of a vine overhanging the terrace; the small slow lizards, brought aboard, transformed the tight-faced distempered tarts into three frightened girls who looked pathetically young and sick in their cheap costumes. Threatened by the lizards, they became hysterical, and their flight in tight skirts across the mess table evoked a good deal of raw laughter. On the deck they tried to pull themselves together, looking injured, then angry, then bewildered.

The ship prepared to sail just after nightfall. Clarkson decided to fly back to Mozambique, where he would check on film locations and see to all arrangements. (Ashore in Mozambique, his health improved, but he never joined the ship again.)

Lindsay had intended to stay with us, but at the last moment he decided to leave with his father: chronic abuse had finally done what seasickness, homesickness and a raw hand had not been able to accomplish. Our villainous cook had taken out on this brave fifteen-year-old the tongue-lashing he had received himself a few minutes before, when he had nearly got the ship detained for another day by sending a bottle of gin through customs on the person of his whore. "I never thought I'd see that boy in tears," Gimbel said, cursing the Dutchman, whom he had nearly struck outside the customs house. Lindsay's spirit had been terrific, and we would miss him; what a difference a year had made in this boy who at Nassau, only last summer, was a child.

At dawn on the 10th, the *Terrier* rounded the south point of Grande Comore, where the volcano called Kathala rose six thousand feet into heavy rain clouds; below the clouds, all around the sea horizon, the encircling sky was radiant.

Moroni, the capital of Grande Comore, is an old gray-white Moslem port with arched passages, small minarets, and carved wood doorways studded with brass knobs. Discovered originally by the Portuguese navigators who named Astove and Aldabra, the Comoros were colonized by Moslems from Zanzibar and Arabia, and subsequently became a French protectorate. Rising onto the mountainside behind the town are the copra plantations of the island, which is best known as the place where a specimen of the great lobe-finned fish called the coelacanth, thought to have gone extinct three hundred thousand years ago, was first obtained.

Fishermen on the quai reported that many large sharks infested the Récif Vailheu, an underwater plateau twelve miles southwest of Grande Comore; we went straight there, arriving about noon. A reconnaissance by dinghy showed that the mean depth of the sea mount was approximately sixty feet, but Jan Moen and Captain Knut approached it with the greatest suspicion: they strongly resisted the idea that an ocean-going vessel should anchor in a place which caused Jan to cry in consternation, "I can *see* it! I can *see* the bottom!" Their emotion was understandable – "But it's our *future*!" Jan had pleaded, off Aldabra. After taking his own sounding with a lead line, Jan signaled to Sam and Francis to drop the anchor, smiling sheepishly when he received a resounding cheer.

At lunch the talk was muted and intense. Perhaps because we had never heard of this strange reef until this morning, it had a certain mystery, and no one had the slightest doubt that sharks abounded. The sea was too rough to launch the cages for a drift across the sea mount, and it was decided that Valerie would spear fish to attract sharks while Gimbel, Waterman and Lake handled the cameras; Ron and I were to guard the group with bang sticks. I told Gimbel that he was sending a boy on a man's errand, and he laughed, pretending that I was joking; Tom Chapin asked me to please not get bite marks on his diving suit.

On deck, assembling the pieces of my patched-together outfit, I knew I was still an amateur. This was one of those days that Gimbel talks about, when for no good reason one simply doesn't have it, and in my dread, I had to be reminded how to adjust my regulator, calculate my weights, and rig my harness loops in such a way that tanks and weights could be shed readily in an emergency. A gray scud shrouding the sun filled me with gloom. I was still pulling my act together when we jumped off the deck, like parachutists, into the gray sea. The water felt cold, and I feared the strong currents, and still I was not rigged right: my weight belt was loose, and the tank valve banged my head, and water poured remorselessly into my mask. But the others were already headed down, and in a small panic at being left behind, I yanked myself into some semblance of order and pursued them.

Below, white ghostly ocean sand rose to meet the black silhouettes, and then the sun swelled the sea with light, defining the form of the strange world, and I felt free again, even exultant; I grinned stupidly in the privacy of my mask.

We cruised along the plateau rim, a place of low corals flattened by the currents. The small coral heads and gray debris were densely populated, and the fish were tame; Valerie shot a big black rock hind almost immediately. She let it swim a little on the spear, and in its exertions the blood welled out of it in a shroud, like greenish smoke, and the current carried the blood toward the deep water. The Récif Vailheu is a sea mount with a sandy plateau as a crown; at the southeast edge the fine white sand slides steeply into the abyss like a snowfield falling over a night precipice. Deep ocean at a reef edge is a classic haunt of sharks. and as a second fish was shot and then a third, the heads of the divers turned expectantly toward the somber blue-black world where the sand vanished.

When the tanks ran out, we surfaced to the dinghy, which had had a hard time following our bubbles in the chop and current. Finally Chapin had jumped in with a snorkel and followed us from the surface, while Cody trailed him in the boat. "I'm glad to see you," Tom said when I popped up beside him with my bang stick. "I felt kind of exposed up here." As Lipscomb's assistant, Tom had had no chance to dive himself, but he never complained; his good nature was indomitable.

The dinghy carried some fresh tanks, and one by one we swam down again. The sun had gone under, and on the bottom, peering around the gaunt drowned realms, and seeing only Lake and Taylor, I supposed that Stan had remained on the surface with Valerie and Gimbel. Ron and Peter had started off, and I went after them.

We swam gradually along the reef, but saw no sharks. Ahead, the *Terrier*'s anchor chain was a dim solid in an amorphous world of roiled sand and silted shapes; the fish fled before us into the shades. Far above, the hull was rolling, throwing a heavy surf to either side. Still there was no sign of Stan, and I trailed the others to the surface. As I climbed the ladder, Gimbel came around the stern of the *Terrier* in the dinghy, and my vague uneasiness focused when he yelled, "Is Waterman with you?" I shook my head. Gimbel handed up his cameras and roared away, and I called out to Anson Lloyd and Sam to send a man aloft; when my tank was shed, I ran up to the bridge with a pair of binoculars, condemning myself for having swum off without making sure that Waterman was not in trouble. A single head is very hard to find in a rough sea, and there was a strong current running; a wave of dread went through the ship. But in minutes Stan surfaced at the ladder. Adjusting his gear, he had fallen far behind but had made his way along the reef edge without difficulty, and arriving at the anchor, had stayed on the bottom for a while to watch the fish. Still, it had been bad teamwork. In rough conditions in the open ocean, no diver should ever be alone.

THE EXPEDITION SAILED from the Comoros that same evening in the hope of working the St Lazarus Bank the following day. From St Lazarus, the ship would steam southward along the coast of Mozambique. I had no visa for Mozambique, and my passport was checkered with entry and re-entry stamps of Tanzania, with which the Portuguese colony is engaged in an undeclared war. From the shore crew at Lourenço Marques we had already heard that the Portuguese were even more nervous about subversion than the South Africans, and as I had to be in Hawaii in ten days, I decided not to risk incarceration; I would stay behind in the Comoros.

We had a fine dinner at the Kathala Hotel, and I walked the film crew to the quai. There we were overtaken by three French *plongeurs* who had heard of the presence in the Comoros of some American and Australian *confrères*, but they barely arrived in time to wave farewell. These men were so upset by the departure of Ron Taylor, whom they had read about in diving magazines, that after the ship's sailing they honored me instead, inviting me out for drinks that I did not need.

The three assured me that big sharks occurred on the Récif Vailheu, sharks as big as the sofa that we sat upon, and as many as thirty at once; they themselves had encountered big *marteaux* and *tigres*, which were especially common. But the sharks came in calm weather, following pelagic fish that circled the sea mount when the sand was not roiled by surge and current; one must come in September or October, they said, and not be in such a hurry.

The three knew little of the great white shark, but became excited when I told them what the Taylors had said about the white: how it raised its head out of the water like an orca to look into a boat, and how it was said to attack boats to get at the occupants. One day, they said, some local pirogue fishermen of a village not far from Moroni had been towed out to the reef by motor launch and left there to fish; according to the survivors, a "large sharklike creature" had reared its head out of the water, then attacked the pirogues, in the course of which seven men had drowned or disappeared. Since that day, no fisherman of that village had ever returned to the Récif Vailheu.

The Frenchmen spoke of a gigantic sea bass that was readily seen in a wreck off Juan de Nova, three hundred fifty miles to the southward; I got off a letter to Peter about this fish next morning. They also spoke of a former habit of fishermen in the Comoros of eating coelacanth, a delicacy they can no longer afford now that specimens of this archaic fish bring such a price. That day, two of the kindly *plongeurs*, Jean Nicholas and Charles Pouchet, showed me four of the big black fleshy-finned creatures; they lay in tubs of formalin in a shed near the quai, awaiting shipment to the outside world.

On two days of the week, one can fly out from the Comoros to Dar es Salaam, in Tanzania: I looked for the *Terrier* in what I calculated to be the region of the St Lazarus Bank, which lies a hundred and twenty miles from Grande Comore. At one point the plane passed over a ship that was anchored or lying to – there was no wake – but the ship was half hidden by low scud, and I could not be certain that it was the *Terrier*. Whitecaps were visible, but the sea was not nearly so rough as it had been in the Seychelles, as I could see clearly from the low altitude that Air Comore's two-engined Piper had attained. A Piper is not a very big plane in which to make a 200-mile open-ocean crossing, and I was glad to see Mafia Island and the coast of Africa.

The *Terrier*, arriving in the region of the St Lazarus Bank at daylight on July 11, had located the sea mount by midmorning. Not long thereafter a small plane passed over to the northward. Gimbel thought, I'll bet that's Peter! perhaps at the very moment that, on high, I was thinking the same thing.

A submerged mountain fifty miles off the Mozambique-Tanzania coast, St Lazarus rises a mile or more out of the deeps to an uneven summit or plateau that ranges from twenty to ninety feet below the surface. It has always excited fishermen and divers who have seen it on the charts, but the ocean that rolls over it is rough, and no one has ever dived it. Gimbel had heard that Cousteau had wished to try it in 1968 but was "weathered out" and now the *Terrier* wallowed disagreeably in seas that ran ten to fifteen feet before a 25- to 30-knot wind. "We were weathered out too," Peter wrote later, "but we were desperate, so we dove it anyway."

There was little or no current on the bottom, and big groupers and snappers (cod and hind to the Australians) were present in abundance to bait sharks. But the bank itself, which tapered off quite gradually at the edges, lacked striking contours and was generally a disappointment. Its flat, amorphous surface was studded sparsely with low corals, and there was a broad bed of green algae, but essentially it resembled the Récif Vailheu. A number of fish were speared and slashed, until blood drifted like smoke across the bank, but no sharks came. The ocean sky was dull and the bank was gloomy, and the next day the dull weather continued. But that day two more dives were made on a shallower region of the bank, and in the afternoon the effort was rewarded. The sun came out, bringing the bank to life, and a thick four-foot barracuda with a striking black base to its heavy tail was photographed being fed by hand.

Valerie's description of this creature is beautiful:

"There was a stillness in that great silver fish that made him stand out stark and aloof. Down there everything moved – the water, the sand, the smaller fish, even we moved, but not that barracuda. He just hung there, still as stone, watching and waiting. I had never seen so much power so contained. Even holding against the current seemed effortless. While we clung on with fingers and toes fighting clumsily to retain our position he just watched. A million generations of selective breeding had produced this perfect fish, and he seemed to know it . . .

"The Saint Lazarus bank had paid off, finally really paid off. Barracuda are considered as dangerous as sharks in most parts of the world. What a great sequence it will be in the film having one of these fellows feeding from the hands of the divers. It was almost as good as filming a hammerhead, and will have a great impact on the general public and skin-divers alike."

* * *

On the 13th of July, the *Terrier* took on water in Porto Amélia, a taut little colonial town on the north coast of Mozambique. Here she was met by Philip Clarkson, who flew in with good news about Mozambique Island, farther south: on a reconnaissance flight in a light plane, he had seen big sharks, as well as a good number of manta rays.

In Porto Amélia, a dive made outside the harbor revealed chiefly that the water clarity was poor, and since the port captain had commanded Gimbel not to snoop beneath the sea until permission to do so had come down from higher authorities, the *Terrier* sailed southward on the second day, arriving at Nacala at dawn on the 15th. A lovely day for diving was spent at the fuel dock, and that evening the ship put to sea again, bound for Mozambique Island. There the authorities were nervous because a dive had been made on the way into port, even though this region was well south of the country's borders with hostile Zambia and Tanzania, and the next day the port captain, a bitter, timid man who had been demoted from his former post as governor of the province and did not intend to be demoted further, refused the crew permission to enter the water, much less take photographs, as a precaution against undersea subversion. Subsequently the ship carried a green-shirted security policeman on every trip outside the harbor.

That night Gimbel flew to the capital, Lourenço Marques, eight hundred miles south, to try to straighten matters out. There he was told that permission to film the underwater territories of Mozambique would have to come from the Foreign Ministry in Lisbon, and he went to the U.S. Consulate for help. The American authorities were more than obliging, and sent off batteries of cables, but meanwhile the *Terrier* lay idle.

"We are presently in the doldrums," Stan Waterman wrote from Nacala, where Gimbel had cabled that the ship should proceed in his absence. "This is the first colony established by the western world in eastern Africa. It is also the last to survive. The fact that in more than five hundred years this colony has achieved no greater affluence than it has today will tell you something of the stultifying effect the Portuguese mind and government has had here. We soon discovered that special permission is required to film along the coast. It seems that Zambian agents may lurk like compressed genies in our aqualung tanks. No official is willing to take a chance and make a decision in this matter. So we vegetated in the harbor at Mozambique Island while Peter flew to Lourenço Marques to muster the forces of good in our behalf. Now coded cables have gone from our consulate to the U.S. Embassy in Lisbon, and the ether is filled with urgent messages to Washington, CBS, and others."

On the way north to Nacala, a big bull sperm whale had been sighted, lazing on the surface. An attempt was made to head him off in the dinghy in the hope of obtaining some underwater film that might be cut into the footage on the whaling grounds, but the beast proved just as obdurate and flighty as the bureaucrats in Mozambique. "Poor Peter, I feel for him, I really do," Valerie noted. "We sure are having some rotten luck."

To add to Peter's frustrations, word had come that Durban's beaches had been closed, due to the shark hordes that were pursuing great shoals of mackerel close inshore; the nets off the beaches were catching as many as seventy-two sharks in a single day. In Natal it was the southern winter, and the water was colder, and among the species in the nets was the great white shark.

A letter from Peter:

"I know that I was going into a decline at the time you were with us at Astove and Aldabra: the consistent absence of sharks and other big fish combined with the shortness of time remaining in the schedule, our budget position, and certain harassing communications from the studio were grinding me down. It got worse after you left, and I remember once at midday, after a particularly uneventful dive, plunking down heavily on one of the banquettes in the messroom where Valerie was having a cup of tea alone. 'I feel drained out,' I told her. 'I've run out of gas.' 'Yes,' she said, 'I've been watching it happen and it hurts me very much.' About that time, I remember wandering into Jim's cabin, sitting down with my head literally in my hands and saying, 'Is there any way we can still salvage this thing?'

"For the first time since before we left on location, I seriously considered the possibility of failure; not just the chance of a mediocre rather than an excellent outcome, but the possibility of a dead loss, going home without an acceptable film.

"It became starkly clear that I had better begin taking pretty seriously your advice to distribute more of the load and to save all my energy for thinking about the film. My discouragement must have been plain to see because as I passed along the responsibility for some of the things I had been attending to, it was grabbed up almost gratefully by Waterman, Lipscomb and Lake.

"Of course Waterman is probably the most generous, gracious person I've ever known. His generosity goes way beyond the material domain; he seems determined to see to it that credit is recognized where it's deserved. But his style is graceful and funny, not heavy-handed, and when anyone tries to give him a gold star he laughs it off.

"Sometimes I've been furious at Stan, and I regret it. My constant fear of losing a vital piece of equipment (much of which was made up specially), and my preoccupation with avoiding any waste of time, have made me thin-skinned and irritable, and I've taken out most of my tension on Stan, who is not as technically or mechanically proficient as Ron or me. Once I asked him to assemble and lubricate all the stainless-steel powerheads. These items were custom fabricated at $40–$50 each, and at the rate we were losing them we would run short long before the end of the trip. But, most important, when we were swimming clear of the cages, they were our only effective weapon against sharks. He returned in about an hour, saying that all the powerheads were ready except one that had given him trouble. The shaft that screws into the handle was scratched and wouldn't pass through the hole at the base of the chamber.

"'That's okay,' I said. 'We'll file the shank smooth and run a die over the threads. It'll be fine.'

"'But I threw it overboard,' he said. 'I figured we had plenty.'

"I think the look of utter disbelief and anger – though, God knows, I tried not to show it – that passed across my face must have been awful. Stan apologized soulfully. Later during the trip when I would ask him to perform some task with a piece of equipment he would sometimes say, 'I promise not to heave it overboard!' But there was never any recrimination in it, nor in anything that he ever said; he never needed to save face."

While awaiting an answer to the cables, the *Terrier* reconnoitered the Mozambique coast and ventured two reconnaissance dives in remote places; the water was murky everywhere, and the fish life sparse. No word from Lisbon arrived at Porto Amélia, and Gimbel could wait no longer. On the night of July 24, after twelve days of pure frustration, he sailed for Grande Comore by way of the St Lazarus Bank. On the sea mount, conditions were much the same as they had been two weeks earlier: the seas were rough and no sharks came to the cut fish, despite a quantity of whale oil dripping steadily over the stern. Even so, an action sequence was obtained.

As usual, the current did not run clear to the bottom; in the last fifteen feet above the sand, there was no current at all. Working in sixty feet of water, in full view of the *Terrier*'s hull, the divers had only to swim along the bottom until they were well up current of the *Terrier* before starting their ascent. But one day Peter Lake, ascending too soon because he ran out of air, missed the ship and was carried off the edge of the bank into the open sea. His bobbing head was spotted by Sam Lloyd, who also had spotted lost divers ten days before, when the dinghy lost track of them in choppy seas off Porto Amélia. Since no boat was ready, Gimbel and Waterman dove into the water and swam toward the drifting Lake; for all they knew, he was in trouble, and in any case he was in the oil slick, which was there to attract sharks.

In her diary Valerie wrote, "Peter Lake, probably due to his lack of experience, was swept away in the strong current and had to be, with much drama, rescued. He didn't seem to mind; but I worried a bit as we had whale oil dripping and he was in the slick. Peter and Stan, to my amazement, both leapt over to save him and had to be rescued also. Peter L. had about 500 yards' start on them and they never even saw the person they were trying to rescue. Peter L. was wearing his vest so was quite okay. Ron, Anson, and Sam lowered the dinghy and picked the three of them up.

"Previously Stan and I had a drama of our own. Stan surfaced too soon, as his air ran out. I looked up and saw him swimming like mad but going backwards anyway. I still had 500 lbs. The safety rope was somehow wrapped around the hull of the ship. I moved up current and headed for the surface, only just managing to get the rope which I dragged over to Stan. He then pulled himself in only to find no one there to take his camera. We both crashed up and down bellowing for help. Somehow the rope trapped one of my legs against the hull of the ship and I became stuck fast, hanging onto the lowest rail and rising up and down with tremendous ferocity. Meanwhile Clive, the new messboy, came and helped Stan. Lars, the chief engineer, came to me but it was some struggle, and I couldn't be moved until Peter Gimbel surfaced and freed my leg. Jim Lipscomb filmed the whole thing, never once lending a hand or calling for someone to help. It would have been all the same if one of us was drowning. Poor Stan must have been terribly exhausted after his desperate swim without air, and once he was on the ladder, to receive no help with his camera for over a minute, I thought was a bit much."

A resentment of Lipscomb's single-mindedness at the viewfinder seeps out here and elsewhere in Valerie's diary, and was shared by other members of the crew: at times we tended to forget that documentary film-making has to be a dispassionate job, that as a professional cameraman, Jim must be aggressive or he wouldn't be any good. Operating under difficult conditions, he felt obliged to conserve his energy and attention for his own work, which was made all the more exhausting by chronic seasickness. And finally, he was encouraged in his attitude by Gimbel, whose relationship with Lipscomb was excellent. Before the expedition left New York, Jim had said to Peter, "If somebody gets hurt or something, the camera doesn't stop; I'm going to keep right on rolling." And Peter had said, "That's just the way I want it."

Valerie: "Today they reenacted the drama of Peter L. being swept away. All cameras jammed including Jim's. Things just get worse and worse. I feel this showing Peter L. being swept away and Peter G. getting the bends makes us look like amateurs but am keeping this to myself, as we need some excitement in the film.

"We are leaving here tonight and going to the Comoros again. Peter Matthiessen wrote Peter G. some interesting news about the place. Anyhow, we haven't much choice. Mozambique is out of bounds, so we must try other places. Peter G. must be nearly out of his mind."

That the divers were not amateurs is evident from the fact that no one was ever seriously hurt; there were close calls, and there was luck, but luck alone would not have carried them for five months of dangerous work. In any case, Gimbel had no fear of filming the mistakes. Lake's emergency, he felt, would make an exciting sequence:

"Lake is not a careless diver. For example, he is the only person who has never gone into the water on any occasion without an inflatable jacket and it did him yeoman's duty then and on one other dive off Porto Amélia. He didn't lose his poise as a result of this incident even temporarily; luckily there were no big oceanic sharks around. But it's ironic that he got into trouble at this stage of the trip because his work has been so very good since leaving Ceylon, where, after weeks of agonizing . . . I finally told him bluntly that I was unhappy about his performance, that I felt he hadn't been putting out. I think it was painful for both of us; I *know* it was for me. But Lake has the capacity for learning, and from that day – I think it was June 19 – to this, he has been a different man. The diving cylinders are blown up almost before they're dry; the air compressors are well maintained; he has helped me in many thoughtful ways; and – most important – I would bet (I can't know because I haven't seen the results) that his own photography has improved. I have the greatest admiration for the way he reacted to talk that can't have been very pleasant to hear."

On Saturday evening, July 26, the *Terrier* sailed eastward to Moroni, and the next three days were spent mostly at the Récif Vailheu. On Sunday the divers were joined by the French *plongeurs*, Jean Nicholas and Charles Pouchet, who were to depart the following day on a trip to Juan de Nova and Europa; in the shadows of the sea the Frenchmen saw the only shark that the reef produced, but the cameras were nowhere near. On Monday, however, another unforeseen event provided footage.

From Valerie's journal: "I received the biggest fright of my life today. Jan wanted someone to check his anchor which hung over a vertical drop-off. Ron agreed to go down. Peter Lake was supposed to watch from the surface, but for some reason stayed in the dinghy instead and lost Ron's bubbles on the choppy surface. After he had been down ten minutes I started to worry, and after fifteen minutes was completely panicked. Stan, who had gone down to check on Ron, hadn't returned either. Cursing Peter L. ( used a swear word) and cursing Jim, who was filming my distress, I donned a tank and leapt in, only to discover I couldn't descend due to my cold. I swam back and got a snorkel, by this time sobbing with fright. Fortunately Peter L. met me and told me Ron was okay. Peter was very distressed at my upset and was very sweet and comforting. We bobbed around in the choppy seas hugging each other. I was sorry I had been so harsh with him.

"While checking the anchor at around 150 feet Ron had spotted a big grouper, and as we had been having trouble getting chum, he decided to stalk and spear it. This he did without much trouble, only the grouper thundered in under the coral. Stan arrived to see Ron struggling with his fish, so, like any interested spectator, he stayed to watch. Meanwhile, back on the boat, poor me. I was nearly out of my mind. I imagined Ron trapped and mangled under the anchor chain, which was swinging about on the cliff face in a most alarming manner. Stan, of course, must be trying to drag out the corpse. All I could think was that I had lost my husband, my wonderful beautiful husband. It was pretty awful and I have not forgiven him yet for giving me such a fright."

Ron accepted a public chastisement from his wife in silence. When she was done, he remained silent for a while, looking out to sea, and then he said, "What's the matter, Valerie? Do you think I'm some sort of amateur diver who can't be left alone?"

Morale on the ship was not getting any better. Because of the frequent moves from place to place, Peter Lake had not heard from his wife, who was in Nairobi, and his usual sunny manner was interspersed with fits of red-faced rage. Stuart Cody was upset that a night dive at the Récif Vailheu had been aborted by new malfunctions in the lighting system; furthermore, he had burned out two expensive amplifiers in the shark attractors in an attempt to raise the volume. Finally, a pet monkey that he had bought at Porto Amélia had been washed overboard and lost – "a foregone conclusion", according to Valerie.

On the 29th of July, the *Terrier* returned to Nacala to bunker and water. Because no official at Nacala would accept responsibility for being efficient, a whole day was lost taking on fuel and water. There was still no word from Lisbon, and Gimbel wrote off the entire Mozambique location as a disaster. The ship sailed for Juan de Nova on the morning of the 31st.

"The combination of political difficulties in Mozambique plus your note about Juan de Nova may have saved our bacon," Peter wrote in mid-August. "In any case, the combination of the two convinced me that we should clear out of Mozambique and head for the islands in mid-channel. When we reached Juan de Nova, luck finally broke in our favor."

Juan de Nova, a French possession lying three hundred and fifty miles south of Grande Comore, extends east and west for two and a half miles and is half a mile across; forested by mangrove and casuarina and surrounded by steep fringing reef, it is very beautiful. The reef is littered with wrecks, several of which emerge in the fifteen-foot tide, and the celebrated jewfish, estimated by the Frenchmen at six hundred pounds, was said to inhabit a steamship wreck on the northwest reef, in the island's lee. Apparently the fish had moved, because the divers could not find it, nor were they able to get close to a great barracuda, the biggest yet encountered, that was seen in another wreck.

At the southwest point the *Terrier* found an anchorage near the wreck of a freighter, and here the action was immediate. The divers had scarcely begun the spearfishing when they were besieged by small black-tip sharks and by the larger white-tipped reef sharks, of a different family than the oceanic white-tips. The black-tips – striking creatures with a bright-white iris in a wild protruding eye, black pelvic and pectoral fins, and a handsome black border to the tail – were extremely quick, perhaps because their natural prey has access to the innumerable shelters of the reef, and they were also very aggressive. Repeatedly they darted in to seize fish from the divers, and Stan Waterman, for one, seemed much more nervous with these creatures than he had been with the cruising man-eaters off the coast of Natal. When Taylor speared one in the head, the volatile fish shot straight out of the sea and crashed on the shallow corals of the reef.

The water at Juan de Nova was glistening clear, with a striking drop-off as a background, and it fairly swarmed with unicorns, jacks, butterflyfish, and other regal creatures of the reef. Water clarity and a calm sea made it possible to devote two days to continuity shots for the *Hermes* decompression sequence, as well as "tight shots", or close-ups, for the action on the whaling grounds. But Gimbel was especially delighted with the reef-shark sequence, which provided a prelude for the film's climax, whatever that turned out to be.

After midnight on the 5th of August, the *Terrier* sailed three hundred miles due south for another French possession, called Europa, arriving there at daybreak on the 6th. Four miles long and a mile across, Europa enclosed a classic lagoon, but the island was considerably less beautiful than Juan de Nova. Underwater, however, the sea cliffs were even more impressive, and in addition to the reef sharks there were great numbers of big bony fishes, including grouper, snapper, jack and moray eel; a hammerhead was seen, well out of camera range, and also an extraordinary snake-shaped jellyfish twenty feet long. Green turtle abounded: the upper beaches of Europa were pocked with nests of the great sea reptiles that dragged themselves ashore during the night.

Europa's sparse terrestrial life included a visiting meteorologist from Reunion, as well as Jean Nicholas and Charles Pouchet, who were assisting a French underwater film-maker named Jacques Stevens; that first evening, the island's visitors feasted together on barbecued wild goat. Stevens, who was filming the mating and nesting of the turtles, is the man who obtained the pictures of the coelacanth that appeared in *Life*, and he expressed embarrassment about it. Although informed that the fish had been dredged up from its deep-water haunts and was dying in the shallows when the shot was taken, *Life* presented his picture as the first ever taken of a live coelacanth in its native habitat.

In East Africa this past winter, the celebrated *Life* series of a leopard attacking a baboon was a subject of general disgust – the "wild" animals had been cage-starved and drugged, respectively – and the photo story (June 1968) of a "white shark" killing the underwater stunt man of an Italian film crew off Isla Mujeres, in Yucatán, was another phony. It now appears that the shark in *Life*'s photos is not a white shark at all but one of the requiem species, possibly dead (though white sharks doubtless occur off Yucatán: in 1946, on a sand flat not far from Tampico, one was found stranded, apparently choked to death by the human body that protruded from its mouth). Anyway, the port captain, mayor, and doctor in charge of the hospital at Isla Mujeres could find no evidence of the existence, past or present, of the stunt man, much less of the episode itself, which they first learned about in *Life*. A diver friend of Waterman and Gimbel named Al Giddings, who was down the coast at Cozumel when the episode is said to have occurred, heard nothing of it from his friends at Isla Mujeres, and in a detailed exposure of the fraud that appeared in *Skin Diver Magazine* (November 1969) a *Life* spokesman suggests to the author that *Life* had been the victim of a hoax. No such admission was ever made to *Life*'s own readers. Like the coelacanth and leopard articles referred to earlier, the hoax brought considerable financial consolation to its victim.

Underwater at Europa, green turtles were caught and ridden for the cameras – one is shown dragging two divers at once – and moray, jack and grouper were all fed by hand. Personally, I don't care much for this sort of undersea circus. It detracts from the majesty of these creatures which are, after all, the film's protagonists, and it dissipates an imminence of menace which is crucial. For me the green turtle is a mystical animal of great beauty, so that my bias here is clear. Nevertheless I am trying to persuade Gimbel that even as spectacle, a man – much less two men – being towed by a panicked turtle is neither interesting nor aesthetic; it is an obvious dull stunt that underwater film-makers cannot seem to resist. Playing water polo with a porcupinefish, or passing an onrushing shark with a red plastic muleta would be comparably stupid, if not quite so boring.

The cinema audience is not the same as the lecture or even the TV-adventure audience, which seems to respond favorably to constructed fun. But when the playing is spontaneous, as it is so often with Ron, or when it displays the animal without demeaning it, as in the feeding of the great barracuda at St Lazarus Bank, it will come as a welcome rest in the search for the great white shark. What would be fatal would be to slacken the tension of that search, which the audience must not be permitted to forget for more than a few frames.

The fish feeding at Europa may also work, in this case because the fish are so unlovable. One jack snapped Valerie's glove off and consumed it, and the groupers were more persistent than the sharks. Anyone who has ever dealt with a hungry grouper must suspect that a thousand-pounder might be more dangerous than a shark of the same size. Grabbing at Ron's fish, one big one seized his hand and spear-grip into the bargain, and had to be punched away.

"A funny thing," Valerie notes. "You can always tell when the cod [the grouper] is going to take the fish. He sort of goes rock steady, pops his eyes madly, then with a quiver of his tail and a flurry of fins, shoots in as straight and true as a torpedo. Once he has set his mind on a piece of food it is difficult to the point of being impossible to discourage him . . .

"I fed my last big cod for Peter G.'s camera. It was . . . the final touch. He hung off in the deep water until finally, overcome by my tantalizing offer, he thundered in and right in front of the camera gulped down the bait. We did a few close-ups in the deeper water for inter-cuts, said goodbye to our French friends, climbed aboard the *Terrier* and all went home. Home to America. Home to Australia, Home to Africa. *Home*."

A last letter from Gimbel:

August 11, 1969

We have finished and are headed for Durban. It reminds me of the afternoon we left New York four and a half months ago. I had been working terribly long hours for about a month, but in the final week it was pretty much day and night and I was exhausted. There were lots of people around packing, making up lists and checking . . . The . . . hammering and moving of cases and yelling never let up. There was continual noise; most of the crew was to catch a flight to Brussels at 7.00 p.m.

At 3.30 p.m. the racket stopped abruptly and everyone stood around bewildered in the stillness. We were ready to leave.

Now we are bearing 218°, full ahead for Durban, then home, and nobody can quite believe that either. I wander around the ship wondering what I ought to do.

We finished strongly and I feel reasonably content with the results . . . There's no reason to doubt that we have ample footage for a feature movie, and very likely a damned good one. But I am not willing to make anything short of the best film conceivable out of the raw material of the idea: the search for the great white shark . . .

IN NEW YORK in August and September, Gimbel and Lipscomb, with film editor John Maddox, assembled a rough cut, or "assembly", of the footage, designed to prove to Cinema Center that the film needed a climax; the assembly was screened in mid-October. Afterward Gimbel explained to the company executives that the Taylors had invariably located white sharks in South Australia's Spencer Gulf, and that even if his own expedition failed to do, the story of that failure would be more interesting than any conceivable arrangement of the material already obtained. While this was debatable, it was certainly true that the film badly needed the white shark. Since the added expense of an Australia location would be relatively small, it would be folly not to pursue the search to the bitter end.

The film executives, excited by the footage, were mostly sympathetic with Gimbel's request for additional financing and his need for a quick decision: the cages and diving gear had to be shipped off immediately if the expedition was to be on location in January, when the weather was most favorable in Spencer Gulf. But a week later Gimbel was still waiting for a call, and it was nearly November before Cinema Center agreed to advance the necessary sums. The cages and diving gear were sent off immediately, and arrangements for hotel space and boat charters confirmed. Except for Phil Clarkson, who was not in shape to go, and Tom Chapin, who had other commitments, the film crew was still intact. It would leave New York on January 7, flying by way of Sydney and Adelaide to Port Lincoln in South Australia. Weather permitting, it would be on location by the middle of the month.

I traveled separately to Australia, stopping off for a few days in California. In San Francisco I talked to Al Giddings, the diver and diving-equipment designer, and an eye-witness to a white shark attack that occurred in 1963 at the Farallon Islands, twenty-five miles off the Golden Gate.

Giddings runs a scuba shop and diving school called the Bamboo Reef, and one February day, with his friend and partner Leroy French, he led a group of fifteen divers to the Farallons for spearfishing and photography. On this overcast foggy day of winter, the water temperature was about 50 degrees Fahrenheit – the ocean temperature on this coast varies less than five degrees all year round – and the visibility on the bottom, sixty feet down, was poor. Some of the group were already back aboard when French came to the surface. Just ahead of him another diver was towing a string of lingcod and rockfish. There was fish blood in the water, and at the surface the water was roiled and turbid, with visibility less than twenty feet.

Al Giddings was back on board preparing for a second dive when he heard an incredible scream. "Have you ever heard a man scream?" he said, as if hearing that sound all over again. "I mean *really* scream?" He shook his head. "I couldn't even tell it was Leroy, that's how wild it was. But a shark attack never occurred to me. I thought some guy was in trouble with his equipment and had panicked, so I handed my camera to someone else and dove into the water, heading for the noise."

Swimming hard, Giddings lifted his head to get a breath, and saw a sight he is not likely to forget. Thirty feet away, the face of French was screaming at him – "My legs are gone! Help me, Al, don't leave me!" – and at that moment an immense fin rose out of the water, wavering silently behind his partner's head. "We were looking right into each other's eyes, and Leroy knew something awful was about to happen because I went like this" – Giddings let his jaw drop and sat back hard into his chair. "I couldn't help it. That fin looked so big I thought it was an orca." But it was the tail fin of a great white shark that was renewing an attack on French, and French himself, seeing Giddings' face and hearing the heavy thrash in the sea behind him, gazed hopelessly at his friend. A moment later he was dragged beneath the surface, and the sea was still.

"I swam over there," Giddings said, looking numb. "It wasn't a moral decision or anything, I just found myself swimming in a strange kind of dream." Then French popped up again, not five feet away.

Understandably, he was now hysterical, flailing wildly with both arms and crying out in a high moan. To avoid his clutch, Giddings swam around behind him and grabbed him by the manifold that connected his twin tanks. The water was cloudy with blood. Dragging French, he set out on the long slow swim back to the boat.

Meanwhile, the other divers had fled the water, but as Giddings approached the boat, two men jumped in to help. While French was being eased into the cockpit and the others clambered after him, Giddings remained dangling from the foot of the ladder. "I tried to stay cool and direct what was going on, but Leroy's blood was cascading down all over the place, and I guess I was there at least two minutes waiting for that shark. That was an awful long two minutes, believe me."

On its first attack the shark had enclosed French's buttock, calf and hand in a bite that must have been two feet across; the second time it seized him by the foot and ankle. Dragged below the surface, French jabbed the shark furiously with his spear gun, and also released the cartridge in his life vest, which he was wearing that day for the first time in his life. The shark let go.

French said later that his worst moment came when he saw the horror in Al's eyes. His wounds required some five hundred stitches, and he spent four months in the hospital. He is still diving, though not with the abandon of before. "You just don't get over something as bad as that," said Giddings, who needed a while to get over the episode himself. In his opinion, the shark withdrew because of the disturbance in the water, including his own arrival and perhaps even the underwater pop of the life-vest cartridge. "Whites are wary until they are really committed to an attack. Usually they come when everything is very, very still . . ."

Four months after the attack on French, Giddings warned a group of divers to avoid the Farallons, where incoming ships often dumped their garbage before entering the bay. The men disregarded his advice, and that same afternoon he heard on the radio that one of the group, Jack Rochette, had been attacked by a white shark. People in his boat had seen the fin just before Rochette surfaced nearby. Veering, the shark struck him so hard that he was lifted straight out of the water to the level of his waist. But it withdrew when his friends leapt in to help him, and Rochette survived.

Before leaving for Australia, I went out to Point Lobos in a vain attempt to glimpse the migration of gray whales, which move close inshore on the southerly passage to breeding grounds in the Gulf of California. Somewhere in the blowing clouds down toward the southwest, the tanker *Connecticut* lay awash with a cargo of fourteen million tons of fuel, and the seas were so rough after a two-day gale that the sea lions had to fight their way onto Seal Rocks; when they failed to grip the rock on the huge surge, they were washed fifteen or twenty feet back down the foaming ledges.

Walking the steep evergreen headlands, I came to a ledge overlooking a broad gray beach between Point Lobos and the Golden Gate; here, in 1959, Albert Kogler was attacked by a white shark. Like French and Rochette, Kogler had a brave companion, a girl, who drove off the shark by coming to his aid, but the single bite brought on a massive hemorrhage, and he died on the gray beach.

Stan Waterman had passed through San Francisco ahead of me. "The last thing I said to Stan before I put him on the plane," Al Giddings warned me, "was to be careful of the great white shark. It's not like other sharks; you can't play games with it." But Stan has a print of Ron Taylor's short film of a white cannibalizing a hooked companion at the stern of a boat, and he needed no warning: in the film, the ferocity of the attack and the sudden, awesome damage done by the great serrated teeth cause an explosion of blood into the water.

One evening last fall at Waterman's house in Princeton, I saw this film for the first time. Susie Waterman, who watched it with us, seemed resigned to the fact that her husband might be dealing with this beast in January. Her aplomb astonished another guest, who made no effort to conceal her exasperation at the "silly risks" taken by men like Gimbel and Waterman – she found them childish. Learning that Stan had followed Peter out of the cage into the shark packs off Durban, she wanted to know *why* he had done it – what was he trying to *prove*?

"I wasn't trying to prove a thing," Stan said, distressed by her vehemence. "Sometimes there is a job to be done, that's all, and it seemed to me that my job was to stay with Peter." For Stan, old-fashioned obligation and good form are a strong defense against his fear – he has that kind of conscientious courage. As soon as possible, he changed the subject. None of the other three divers, he told us, were ready for Gimbel's sudden exit from the cage off Durban. "It was just one of those impulses Peter gets sometimes," he concluded, looking troubled; he is not the first one to be bothered by Gimbel's hidden drives.

To Gimbel himself, however, there is no mystery. "I have no pride or rules about courage," he says. "I go when I feel dominance over the situation, and not on days when I'm afraid – those are the days that you get hurt." For years he has fought off the suggestion that he is out to test himself, or is ruled by some sort of death wish. One day in the Seychelles, Jim Lipscomb questioned him about the pressure put on him by his powerful and successful father and about his undeniable tendency to place himself in one dangerous situation after another, and Peter responded with the eloquence of a man who has met such questions many times before.

"Danger doesn't interest me," he said in part, ". . . but I'm curious and I think everybody's curious to find out just what their limits are under situations that exert a certain amount of stress on them. I would be just as curious, for example, to know what my limits are as a gambler, but I already know that, so I'm not curious: I'm a lousy gambler . . .

"Look, when we were working with those sharks off Durban we wanted to get as close to the whales as we could . . . Well, what we found out was that there are no limits. We all know now that we could work right next to the mouths of those sharks, and we have the film to show that we could. Why they permitted it I don't know, and it's against all expectations, but it's a fact . . . We couldn't be theoretical about it because all the theory would indicate that it was impossible, but it's a fact.

"You keep asking about this risk factor, but the fact of the matter is that when you get involved with a project like this you do take a certain pride in trying to make the situation just as exciting as it can be. The cages and machines – we know they work, that's not why we came here. We came really, when you get right down to it, to see what the limits are, just how wildly – that's the wrong word – just how openly a man can expose himself in the water with excited sharks and still maintain control."

Peter's arguments are invariably well reasoned and sincere, and yet, sensing that some small piece of self-awareness is missing, one goes away unsatisfied. I have listened to him for years, and I always believe him when he speaks, but still the questions keep occurring. In a careful way, with impeccable preparations, he seeks out ways to test what he calls "the limits", and of course this search has no real end to it but death. Still, I don't think this is a "death wish" unless dread of death is the same thing. It is as if, by confronting death over and over, he might end some awful suspense about it, or dissipate it in some way. More than any man I have ever met, Peter loathes the aging process in himself: "I look in the mirror and I hate what I see there, and it's just happened in the last year," he says, cursing his face lines and gray hair, though in fact his hair turned quite gray several years ago. And this lack of serenity in the face of his own transience seems out of character to the people around him. As Valerie says, "Peter's so beaut the way he is, he shouldn't *need* to suck his tummy in and hide his bald spot when the camera's on him."

Since in some respects our explorations have been similar, I am sympathetic with Peter's need to find out what the limits are; the original motivations may be ambiguous, but attacks upon this life style are often ambiguous as well, as if the need to attack betrayed a fear in the attacker that his own life seeps away from him unlived.

On the plane from Adelaide to Port Lincoln, where I caught up with the expedition on January 13, I mentioned casually to Stan that I had been a sissy as a child. After a long moment and a wary glance, he exclaimed, "Were you really? I was, too!" – as if relieved that such a thing had become admissible. He talked and laughed in great good humor about the indignities he had suffered in the pecking order of school. Later he remarked that for both Gimbel and himself an investment of ego was at stake in the search for the great white shark; neither man is what he calls a "spontaneous adventurer". By this term I assume he meant a man who embraces adventure for its own sake, as opposed to one motivated first by fortune or glory. Certainly Ron Taylor seems much less adventurous than coolly professional, a technician-adventurer, like an astronaut he was hired to do a certain job which he wants to do very well, and if someone calls it an adventure, so much the better. Peter Lake, with his taste for such unmystical pursuits as auto-racing, might qualify as a "thrill-seeker", but he is not an "adventurer" either, however much he may enjoy being funny about being frightened.

In this group, perhaps Valerie comes closest, in the sense that she pursues her goal in headlong fashion. The first day on the Récif Vailheu in the Comoros, Valerie felt so sick that she had to lie down in her cabin while waiting for the dive. When she reappeared, she was still shaky, and I suggested that she take the day off. "What?" she cried, outraged. "And miss out?"

"I'm a lot more adventurous than Ron," she told me in Australia. "Ron's content with his way of life the way it is. If I see a pearl oyster, I want to open it up to see if there's a pearl in it, but Ron can't be bothered – that's the difference between him and me."

In Australia, both Valerie and Ron were much more at ease than they had been on the first part of the expedition. This was true especially of Ron, who had been reserved even with Valerie, and was now quietly demonstrative. With the rest of us, too, while scarcely loquacious, he was spontaneous and relaxed; he had never been distant, as I thought, but merely diffident. In strange countries, as the only Australians in a large group of American extroverts, the Taylors had felt shy; here they were not only on home ground but would be accompanied by other Australian divers. "You should see Ron on the Barrier Reef, when he's in charge of things," Valerie said. "You wouldn't know him. He gives orders and everything!" One day on the Reef, Valerie watched Ron play peek-a-boo with an octopus. "He used up a whole tank, just *playing*! I never knew he had such fun in him! It was terrific!"

"It would seem reasonable to expect," Gimbel says, "that a man so circumspect and reserved might . . . reveal himself in surprising ways. At Aldabra, you recall, we dove the pass leading from the open sea into the lagoon. This channel, only a hundred yards wide and about sixty feet deep, drains or fills the lagoon with each tide. The current builds quickly to better than six knots after the turn of the tide. We entered the water just at slack, but within a few minutes we were being swept along like dandelion pods in a breeze. I caught sight of Ron, who was heavily weighted, traveling through the pass in a series of giant moon leaps. He would spring from the bottom and travel twenty or thirty feet in a soaring arc before sinking again to the bottom. He looked blissful.

"At Europa I filmed him completely surrounded by a school of exquisite blue-and-yellow fish, swarming so thickly that at times he was entirely obscured from view. When the school enveloped him completely, he would push his hand suddenly into their midst, and they would part like a curtain struck by a sudden gust; then almost immediately they would close about him again. He did it to a gentle, playful rhythm that delighted me." Ron's instinct for whimsy rarely showed itself above the surface. People play where they feel confident, and Ron's play was always underwater. In New York I had seen one very funny sequence shot off Porto Amélia in which Stan finds a starfish and pins it on Ron's black wet-suit like a huge medal. Ceremoniously, awed by the honor done him, Ron raises a white-gloved hand in slow-motion salute, clasping his starfish to his heart with his other hand. One looks for the tear in the eye behind the mask as he turns, hand still upon his heart, and swims off in an elderly manner into the blue.

PORT LINCOLN IS located on the barren Eyre Peninsula of South Australia, which forms the western shore of Spencer Gulf. The foremost fishing port of Australia and a shipping center for the wheat and livestock ranches of this region, it is also a summer resort with a beach front on its broad pale bay. South of the town, toward the uninhabited tip of the peninsula, is a dry rolling scrub of gum and casuarina and melaleuca where at dawn one morning we saw kangaroo and emu. Westward, the scrub dies away in the dry wastes of the Nullaboor Plain, the Never-Never of the aborigines. Much massacred in the last century, the aborigines are now scarce here, but a few are to be seen in the local bars. These remnants have taken on the raffish habits of the "Outback" whites, who are finishing them off through interbreeding. Like the American Indian, whose fate so resembles their own, the native Australians had a wealth of legend and symbolic art, which becomes ever more fashionable as its creators die away.

The expedition was housed at the Tasman Hotel, overlooking the beach boulevard and the bay; a storehouse-workshop was set up in a shed behind. Logistics ashore were handled efficiently by John Carey, while those at sea were the province of a champion skin-diver named Rodney Fox, whose experience in the waters of this region includes a near-fatal attack by a white shark.

The expedition was welcomed warmly by the mayor and other officials, who held a press reception in the town hall. There Jim Veitch, an old-timer in Port Lincoln, showed us the jaws of his 15-foot 9-inch specimen, weighing more than a ton, which was once the world-record white shark caught on rod and reel; the jaws fit easily over Gimbel's head and down over his shoulders. "Beautiful set of teeth," said Rodney Fox, who still carries a beautiful set of teeth marks.

The hospitality extended by the town included access to the town tennis courts. The search for the great white shark had been delayed by windy weather, and Lipscomb, Waterman and I played tennis every day that the expedition was in port. Meanwhile, equipment was uncrated and sorted, and provisions loaded aboard the *Saori*, a motor ketch chartered for the work at sea. Designed for seismic research, the ketch was a clean, able double-ender, sixty-five feet long, with a boom forward to swing the cages into the sea. Her deckhouse was given over to cameras and sound equipment, and the film crew would share the bunks and decks of captain's cabin and the galley-fo'c's'le. Waterman was delighted with the fo'c's'le, which inspired him to heady speech of scurvy Lascars, pots of rum, sea shanties, dicing, tales of wickedness in port, and other nautical extravagances.

The 18th of January. We are bound south for Cape Catastrophe, which guards the mouth of Spencer Gulf; from the cape westward, an inhospitable and scarcely inhabited desert coast stretches for a thousand miles to the southwest corner of the continent at Flinders Bay. The Australian divers on the *Saori* think that Spencer Gulf is the best place in the world to film the great white shark, which is drawn inshore by sea lions and by the fish and fisheries based at Port Lincoln. Besides the Taylors and Rodney Fox, these divers include Ian McKechnie, a one-time abalone diver, tuna fisherman and kangaroo hunter, who replaces Tom Chapin as Jim Lipscomb's assistant, and Bruce Farley, the cook-deckhand on the *Saori* who, like Fox, is a former skin-diving champion of South Australia.

At dawn, Spencer Gulf is as still as mirror silvering, and a pale sunrise casts a still, pale light on the shores of its lost islands. Rounding Cape Donington, the *Saori* meets the cold southwest wind that blows up from the Antarctic Ocean; steered by Ben Ranford, an imposing old sea dog with great grizzled eyebrows, white hair on his nose, and a kind quiet face, she moves southward under the lee shore. Attending the *Saori* is a Diesel powerboat, the *Sea Raider*, which will be used for auxiliary jobs and fast transport in and out of port.

Gimbel makes no secret of the fact that the auxiliary jobs he has in mind could include emergency transport of a white shark victim, who might not survive a slow trip on the *Saori*. One night at the hotel he had said to Lipscomb, "Let's see now, who'll be on deck to do first aid in case of trouble? If we eliminate the people diving, that leaves only Stuart and Ian McKechnie and Rodney –"

"That means Rodney," Lipscomb said. "Stuart and Ian have to stick with me. And you'd better remind Stuart that his job is to get that sound – he might get humanitarian on us at the last minute."

Gimbel had selected Cody as the man to inject a painkiller and shock inhibitor called Demerol. He had even extended his own arm for Stuart to practice on – "a beautiful illustration," Cody says, "of the thoroughness of his preparations and also of that sense of drama that is just right for this kind of operation." But Gimbel had nodded at Lipscomb; he apparently agreed.

Sitting in the corner with my drink, I shifted in discomfort – how would we feel afterward if a humanitarian impulse was repressed in favor of a strong sequence for a commercial movie? I raised my eyes to find Jim watching me. Doubtless he knew that I would record this conversation, just as he intended to record any agony on deck.

Memory Cove, backed by headlands of the Eyre Peninsula, lies just inside Cape Catastrophe. It was discovered in 1802 by the great navigator of Australia, Captain Matthew Flinders, and, like the cape, is named in memory of six sailors of H.M.S. *Investigator* who were drowned here when rough seas capsized their boat. In the heart of the cove is a still beach, as white as the curved rib of a whale. From Memory Cove, the northeast horizon is broken by bare islets of Spencer Gulf; due eastward looms a high dark mass called Thistle Island. The cape is only 5 degrees north of the rude latitudes called the roaring forties, and incoming seas backed by the prevailing winds explode upward against steep seaward cliffs of the dark island. Nowhere is there sign of boat or human habitation: it is like an ocean on the moon. A gull drifts down the cold clear sky, and on a shining rock at the point of the cove, against the blue of the breaking sea beyond, a dark unknown bird faces inland.

The *Saori* had scarcely anchored when Rodney Fox filled a drip can with whale oil and hung it over the side; the slick, blown seaward, would spread into the gulf. Nearby he suspended four Australian salmon (not a true salmon) to keep incoming sharks around the boat. On two trips out of three, Rodney has raised white sharks at Memory Cove, and Ian Wedd, the skipper of the *Sea Raider*, saw one here just a year ago. "Nearly lost me daughter," he said cheerfully. The girl was playing in the turquoise margin at the edge of the beach when a big shadowy shape came gliding in from the dark water. Ian called to the child, who splashed ashore, and the shark followed her into the shallows. "A fair dinkum white death, it was," Wedd said. "About a ten-footer, I should say."

"I rarely swim where whites are," Valerie commented, "and if I do, I have a chlorophyll tablet first." Chlorophyll tablets temporarily suppress one's natural odor, and Valerie is convinced that the three friends of the Taylors who were attacked by the white shark share a distinctive smell – "not body odor, but a bird-cagey smell." She glanced at Ron, who didn't deny it. "I *know*," she continued, "because I've had all three in my spare room and had to air it. And only one of the three chaps doesn't wash – it isn't *that*." Two years before at Memory Cove, Ron, suspended in a cage, failed to get decent film of a white shark because the shark would come no closer than thirty feet. A little later, when his friend Henri Bource entered the cage, the shark came straight up and nosed the bars. "We like diving with those three," Valerie said. She laughed, but she was serious, and her husband nodded.

The year before, Henri Bource had been diving off the seal colony at the Lady Julia Percy Islands, in Victoria, when suddenly the seals vanished, leaving only trails of bubbles and a great stillness. A moment later a white shark seized his leg and dragged him forty feet down. Bource was wearing a black suit, and as white sharks are thought to cull decrepit animals out of the seal herds, Bource may have been taken for a seal that could no longer get away. In any case, he would have died had the shark's bite been less powerful: "To my great relief" – those were his very words, Valerie says – "I felt my leg give way at the knee." That white sharks can catch seals at all is a testimony to their speed. Henri Bource's experience reminded me that in no case I have ever heard of has its victim ever seen the white shark coming.

Possibly Valerie's odor theory helps explain a phenomenon that has often been noted in shark attacks: once it has singled out its prey, the shark will usually persist in attacking it to the complete exclusion of other targets, so that rescuers coming to the victim's aid may be scraped and buffeted as the animal returns to the attack but are very rarely bitten. On the other hand, a shark often retires after a single bite, and Ron and Rodney think that this may be because human beings do not react as shark prey should, and the shark gets uneasy. "They're just after a feed," Rodney says, "and this isn't what they're expecting."

The whale-oil slick had drifted a mile or more up the blue gulf, and the crew settled down to wait. The cages, corroded by use in the Indian Ocean, were checked out in the water; both had air leaks in their flotation systems which were quickly fixed. Cameras were given final adjustments, tanks were filled with compressed air, and new powerheads were assembled. Each cage was fitted with two powerheads in case a shark tried to dismantle it: Ron Taylor, for one, has no doubt at all that a white shark could bite through the aluminum bars.

In midmorning a seal broke the surface of the cove, thrashing and playing off the bows; it was an eared seal, the South Australian sea lion. A crested tern, bright-white in the blue sun, appeared and vanished. From time to time, small sprays of bait fish showered the surface, but no shark came. There was only a great stillness, and sere islands in a metal sea. In the late afternoon, when underwater photography was no longer possible, cages and cameras were taken to the bottom on a test run; the water temperature was 60 degrees, and before long the divers were shaking with cold. Back on deck, Waterman called down to Gimbel that he should swim from the cage to the ladder instead of trying to haul himself aboard. "I'm not swimming here," Gimbel said shortly, clambering up the ship's side, tank and all.

At twilight a small party went ashore to explore the mallee scrub behind the beach, and with Jim Lipscomb I walked inland across the peninsula to gaze from the sea cliffs at what Australians call the Southern Ocean. Thirty miles to the south, out of the wind haze, rose the Neptune Islands, and beyond lay the Antarctic Ocean, which was, in my travels, the last of the Seven Seas. Brown swallows called fairy martins crossed from headland to headland, and below, big black-backed Pacific gulls came sweeping in across the broad wash of the reef. On this lonely coast, with the wind from the south out of Antarctica, it occurred to me that the air we breathed might be the last clean air on earth.

During the night I hoisted the lines to which the dead fish had been strung; they were untouched. The whale oil, which sometimes congeals when the air is cold, was still dripping silently into the sea, and I refilled the container. Overhead, the southern stars were round and bright in a ringing sky: Canis Major and Canis Minor, Castor and Pollux, the Southern Cross, Orion. Orion's belt is the leg of a mythical South American Indian, Nohi-Abassi, whose wife had him bitten to pieces by a shark for having persuaded another shark to devour his mother-in-law; the pieces were scattered all across the heavens. Shark legends occur around the world; the Jonah of the North Australian aborigines is Mutuk, who was swallowed entire, and some say that "the fishe" that swallowed the Biblical Jonah was the great white shark.

At daybreak the water was as still as lead – no sign of life, no bird. The skiff crossed Thorny Channel to the islands. In the distance, on the ocean points, the black rocks shone, and the still bodies of the sleeping sea mammals were brilliant, reflecting the dawn like things being given life. On the nearest islet the sunrise spun on the white heads of gulls and turned the amorphous forms of sleeping sea lions to gold. Golden animals lay on bronze carpets of algae, and glistened in the white surge of the shore. They reared high to see into the skiff, and one flashed like a porpoise through the shadows of the morning sea, curving away under the hull. Near a strange balanced rock, upright as a monolith at the water's edge, two young gulls haggled, yanking at the downy body of a bird still younger.

To bait any sharks drawn to the colony, we ran a whale oil slick around the islets, then back across the channel to Memory Cove; later in the day the *Sea Raider* ran a slick out to Cape Catastrophe. In the silence, porpoises sliced softly through the waters of the cove, and gulls came to the fish scraps in the slick, but the white shark did not appear. After forty-eight hours the *Saori* returned to port, passing the cove under Cape Donington where Jim Veitch had caught his monster shark.

According to Veitch, who has studied them for forty years, the largest white sharks caught in South Australia have been found close inshore, and all were females; sharks taken at Dangerous Reef in the middle of

Spencer Gulf are almost invariably males. This led Veitch to suppose that the females come inshore to drop their young, which might be eaten by the males in the open water, but this sensible theory had to be discarded, since none of the females contained embryos or young. Its reproductive cycle is one of many mysteries surrounding the white shark; few embryos or young white sharks have ever been recorded. The smallest white shark Jim Veitch has ever seen was five feet long, and he commented that he would rather catch an 800-pound female with eggs or unborn young than an 8000-pound world record for rod and reel.

Though larger specimens are regularly reported (Cousteau saw a white shark in the Azores that he estimated at twenty-five feet), Veitch has strong doubts about unmeasured whites over eighteen feet long. In fact, he has made a standing offer to *eat* that part of any specimen exceeding eighteen feet, starting at either end. A hooked shark alongside his boat at Streaky Bay, carefully estimated at just over seventeen feet before it got away, was the only one he has ever seen over sixteen, and his record fish was one of the few over fifteen. (A sixteen-footer was the largest of the 141 white sharks that have been netted since 1964 along the coast of Natal.) In Veitch's opinion, a fourteen-footer is a "good big shark". According to museum measurements of the bite wounds, the one that hit Rodney Fox "like a submarine" and bore him away through the water was a mere ten-footer.

Fox, a fair-haired jaunty man of thirty, has been involved in more white shark attacks than any living person, and the pattern of his experiences is weird. In 1961, when he won the South Australian Skin-Diving and Spearfishing Championship – this is free diving, without aid of scuba tanks – his chief competitor was his friend Brian Rodger, who had been state champion the year before. Late one afternoon of March, during a "comp" at Aldinga Beach, south of Adelaide, the two were swimming close to each other when a shark seized Rodger by the leg. Rodger wrenched himself free, but the shark came in again, and this time he deflected it with a point-blank shot from his sling spear gun. Though the barb scarcely dented its tough hide, the shark veered off, and Rodger, bleeding badly – his wounds required some two hundred stitches – used the rubber sling from his spear gun as a tourniquet on his leg, then struggled on, unaided; he was finally picked up by a rowboat near the shore.

Fox was beneath the surface during the attack and was never aware of it; all he saw was the swift approach of a white shark that came in and circled him closely, so closely at times that he could have touched it with his spear gun. Even as he spun desperately in the water, he had to keep going to the surface to get air. Then he would dive for the bottom, thirty feet down, seeking protection, and creep a little way inshore. Relentlessly the big shark circled, and Fox is convinced that this was the one that Rodger had driven off, returning now along the trail of Rodger's blood; since both men wore black suits, it might have mistaken Fox for its original prey. This distraction, which Rodney thinks could not have been less than ten minutes, may well have spared his bleeding friend from further attack.

As the minutes passed and the shark persisted, Rodney had to fight a growing panic. He was still a half mile offshore, and was spending his last energies going to the bottom. Even when the shark was gone, he felt certain that it would return, and the day was growing late; he was most frightened of all that dark would fall while he was still alone in the open water. But the shark never reappeared, and he got ashore.

Fox became state champion that year and was runner-up the next; in 1963 it was expected that he would regain the South Australian title, and Ron Taylor, who was champion of New South Wales, thought that Rodney was the man to beat for the championship of all Australia. Once again the South Australian competition was held at Aldinga Beach, which is noted for its plentiful fish and is only thirty-four miles south of Adelaide, and this time Fox was swimming near Bruce Farley, whom he had taught to dive. That Sunday there were forty divers in the competition, which was based on the number of fish species taken as well as total weight; all contestants wore black wet-suits, and all dragged their fish behind them in a plastic float to minimize the amount of blood in the water.

By early afternoon, when he started his final swim, Fox appeared to be well ahead. On his last trip to the beach with a load of fish, he had noticed two large dusky morwongs near a triangular coral head, about three quarters of a mile offshore. Returning to this place, he parted company with Farley. "He went one way and I went the other," Bruce recalled, making a diving flip with his hand, "and the next thing I knew, the shark had him."

One of the big morwongs was out in the open in a patch of brown algae, and Fox was gliding in on it, intent, spear gun extended like an antenna, when he felt himself overtaken by a strange stillness in the water, a suspension of sound and motion, as if all the creatures of the reef had paused to watch him. "It was just a *feeling*," he says. "I didn't tense up or anything – I didn't have time to." For at that moment he was struck so hard on his left side that his face mask was knocked off and his spear gun sent spinning from his hand, and he found himself swirled swiftly through the water by something that enclosed him from the left shoulder to the waist. A great pressure made his insides feel as if they had been forced toward his right side – he seemed to be choking, and he could not move. Upside down in the creature's mouth, he was being rushed through the water, and only now did he make out the stroke of a great shark's powerful tail. He was groping wildly, trying to gouge its eyes, when inexplicably, of its own accord, the shark let go.

Out of breath, pushing frantically to shove himself away, Rodney jammed his arm straight into its mouth. For the first time he felt pain, a pain that became terrible as he yanked the flesh and veins and tendons out through the back-curved teeth. He fought his way to the surface and grabbed a great ragged breath, but the shark was right behind him. When his knees brushed its body, he clasped it with arms and legs to avoid the jaws, and the beast took him to the bottom, scraping him against the rocks. Once more he fled for the surface, and again the shark followed him up: his moment of utmost horror came when through his blurred vision he saw the great conical head rising toward him out of the pink cloud of his own blood. Hopelessly, he kicked at it, and the flipper skidded off its hide. At the last second the head veered toward his float, which contained a solitary small fish, and a moment later the float raced off across the surface; either the shark had seized the float or had gotten entangled in the line.

Once again he found himself being dragged through the water; already he was far below the surface. He tried to release the weight belt to which his float line was attached, but his arms did not work, nor his mutilated hands. And at this moment, when he knew finally that he was lost – "I had done all I could, and now I was finished" – and was on the point of drowning, the next event occurred in the series of miracles that were to save his life. Presumably the shark's razor teeth had frayed the heavy line that connected the fish float to his weight belt, for at this ultimate moment it parted. For the third time he reached the surface, and this time he screamed, "Shark!" There was no need of it; a boat which had brought a young diver from the beach was only a few yards away. "They thought that lad would be safe with Fox and Farley," Bruce Farley said. "They'd hardly dropped him in the water when they had to yank him out again, because there was Rodney, screaming, in a pool of blood. They hauled out Rodney, then came for me, and we headed for shore."

The bones had been bared on Rodney's right arm and hand – his hand alone required ninety-four stitches – and his rib cage, lungs and upper stomach lay exposed. "Bruce thought I was done for," Rodney said. "The rotten dog sat up in the bow with his back to me – wouldn't even look at me."

Farley grinned. "I just didn't like the looks of all them guts hangin' out," he said. In the boat there was nothing he could do for Rodney, and he tried to concentrate on how best to find help on the beach. "I knew everything had to go one-two-three if we were going to save him, and I didn't even know how bad he was. Oh, there was a little bit of intestine stickin' out, but we never opened his suit up to really see. We made that mistake on the beach with Brian Rodger, and his leg fell all apart." Fox himself feels that his suit, holding his body together until it could be reassembled, was one of the many things that saved his life.

The first person that Bruce met as he ran down the beach was a policeman who knew just where to telephone and what numbers to call. And someone had happened to bring a car down the rough cliff track to the beach – a very rare occurrence; this car was able to bump out onto the reef to pick up Rodney, and it carried him back up the cliff to the highway and eight miles down the road toward Adelaide before he was transferred to the ambulance sent to fetch him. Already the police were manning every intersection on the way, and because he was traveling just before the Sunday afternoon rush he actually reached the hospital within an hour after he was picked up in the boat. His lung was punctured, he was rasping and choking, and that he did not drown in his own blood or bleed to death within that hour was miraculous. Nor were the miracles over: the surgeon on emergency duty that day had just returned from England, where he had taken special training in chest operations.

While Rodney was being prepared for the four-hour operation, he heard urgent voices. One said that someone should go for a priest, and Rodney realized that they thought he was unconscious and did not believe that he was going to make it. Desperate, he half sat up on the table, saying, "I'm a Protestant!" before they got to him and calmed him down. "He's a bloody mess," the doctor told Rodney's wife after the operation, 'but he's going to be all right."

Two reasons for Fox's survival were his excellent condition and the fact that he never went into shock. "It's shock that kills most people in a shark attack," Ron Taylor says, and Valerie agrees. Experienced divers are more apt to survive an attack because they are less apt to go into shock; sharks are a reality that they must live with, and therefore they are psychologically prepared.

"I guess I just wasn't supposed to go," Rodney says cockily. After two weeks he was home in bed, though he had to pay daily visits to the hospital. Six months later he made himself dive again, and he has been diving ever since. In 1964 Ron Taylor's team in the Australian championships was beaten by the team of Brian Rodger, Bruce Farley and Rodney Fox.

The 1963 attack on Fox occurred only a few hundred feet from the place where Brian Rodger had been attacked; forty divers were in the water on both days, and on both, it was the reigning champion who was hit. In 1964 Bruce Farley was state champion. One competition day he was to drive down to Aldinga with Rodger and Fox, but somehow got left behind. By himself, Bruce drove five miles out of town, then turned around and went home. "I can't account for it," he says. "I just lost interest."

That same day at Aldinga, a year to the day after the attack on Fox, both Rodney and Brian separately, simultaneously and for no good reason started for shore. The competition had another hour to run, and both men habitually stuck it out to the very end, but today they each had an instinct to leave the water. Perhaps the two had heard that stillness which precedes the coming of the white death, because before they had reached shore someone came yelling down the beach. A young diver named Geoff Corner had been bitten just once, on the upper leg, but the great bite had severed an artery and he had died. Geoff Corner was the reigning junior champion.

Since that time, Bruce has never dived Aldinga Beach. "We're nice to Bruce," Rodney Fox says, teasing him. "We always decide we'll go and dive somewhere else, because he can't dive Aldinga with his heart and soul."

Bruce Farley is an honest man with a sad humorous bony face. "I haven't dived *anywhere* heart and soul," he says, "since Brian got hit in 1960."

Valerie thinks that the same animal was involved in the series of attacks off Aldinga Beach. In any case, the records indicate that the species is more common in Spencer Gulf than at Aldinga. Everyone is confident that white sharks will be encountered here, but whether they can be photographed is another matter. Many whites are extremely wary, and compressed air bubbles are said to put them off; the divers searching for abalone in these waters use compressed air piped from their boats, and in Australia there is no recorded case of a white shark attack upon an "ab diver".

Like most ab divers, Ian McKechnie has never laid eyes on a white shark, though once one came and took a seal close by. According to Ian's mates, who were yelling to him from the boat, the seal had been swimming only twenty yards behind him. Rodney says that local fishermen in small boats fear the white death and its sinister habit of rearing its head out of the water; when a big one comes near, he says, they crouch below the gunwales to avoid being seen, then start up the motor and escape.

At dawn on January 21 the *Saori* put to sea again, bound this time for Dangerous Reef, an east–west litter of reefs and low rock islets where

Rodney once saw seven whites together. With the sunrise, a strong northeast breeze increased to a battering wind, and there was no lee at the reef, which is marked by a solitary light. The ketch trudged west across the gulf to a lee behind Taylor Island, where a sheep bought from the local slaughterhouse had its throat slit and was strung over the side; the smoke of sheep blood in the water drifted down the shore. The wind shifted to the south, then backed around to the northeast again during the night, but no shark came.

By morning the wind was easterly, at 10 to 15 knots; it slacked off after sunrise, then picked up again at midday. The *Sea Raider* reconnoitered Dangerous Reef, and a party went ashore on its main islet while awaiting the *Saori*. The skiff entered a rock pool in the broad granite ledge used by the seal pups as a nursery, and the lion-colored cow seals, howling, hurtled past the skiff into the water, then reared up all around it, trying to frighten it away. Everywhere rushed new seal pups and small yearlings, fat and astonished, and scattered through the herds were harem bulls with a golden nape on their big neck muscles and dark brown coats which on the older bulls looked black. One cow rushed up out of the water as her pup was approached, barking aggressively and flopping with great agility from boulder to boulder; when faced, she stopped, swaying her head and neck in consternation. Up and down the rocks the young pups shrilled, adding their voices to a din of life that from a distance sounds like one mourning cry.

On the rocks, facing upwind, stood companies of Pacific gulls, a sooty oystercatcher, and small bands of sandpipers and turnstones, picking crustaceans in the rich algal mats that covered the granite platforms washed by tide. Offshore a sooty shearwater pointed its long wing at the water, stiff as a boomerang. In the high sea wind and cold austral light, the black-faced cormorant, stark black and white, came up in strings out of the south; other cormorants stood snow-breasted on the white-stained boulders. Their nest, of a lavender kelp and a brown rockweed stolen freely by the birds from unwary neighbors, held two or three eggs of a watery sea-green blue, though here and there a brown egg lay beside a blue one. The reptilian birds croaked dismally, hissing at the pirate gulls; as fast as a cormorant left its nest, the gulls fell upon its eggs with a horrid squalling. On the high ground under the light was the islet's only growth, a small succulent burned by guano and beaten flat by salt and wind.

At nightfall the wind was still out of the east, but the weather report predicted a shift to the southwest. Since any such shift might put the *Saori* on the reef, Captain Ben decreed that his ship would go in under Cape Donington for the night. By now everyone was so anxious to get to work that the decision was accepted with the greatest reluctance. "You can't go against the *captain* of the vessel," Ron Taylor exclaimed, dismayed by the mutinous mutterings; it was one of the rare times that Ron ever offered a strong opinion.

Before sailing, a four-gallon kerosene can full of chum and whale oil was anchored to the bottom to attract any passing sharks into the area. Next morning the can was gone. Possibly a big sea lion had taken it, parting the buoy line by force, but the sharp cut on the line – the rope strands had not unraveled as they do when parted under strain – suggested the sharp teeth of a shark.

As predicted, the wind had come around to the southwest, but there was a lee on the far side of the islet. The *Saori* anchored close to shore, scarcely more than a hundred yards from the sea-smoothed beaches, and for the next three days, alone in the lunar emptiness of Spencer Gulf, she was washed by a cool southwest wind, seal moaning and the parched stink of guano.

Day after day, sky and sea were the same oily gray, and the sea glimmered dimly like the old seal-polished stones. Rips and the low spine of rocks at the east end of the islet parted a high chop so treacherous that the surf seemed to fly in all directions, a true maelstrom. With the mourning of the sea mammals and the hurtling black birds, the air of this bare reef carried an infinite foreboding that boded well for the coming of the great white shark. And still it did not come.

On the foredeck, Jim taped Rodney's harrowing account and photographed his scars. The divers sewed hoods onto neoprene surfers' vests to be worn under our wet-suits as extra protection against cold, but otherwise there was little to do except watch for sharks in the empty slick that trailed away into the northward. That afternoon, for want of a better plan, Ron, Stan and Peter ran another check on cameras and cages, and I went with them; we took the cages to the bottom, thirty feet down. The cold sea was full of long pale shreds of a drifting algal bloom, and visibility was no better than thirty feet. The bottom itself was strange and dead, a coarse bed of myriad bits of crumbled coral-like gravel haired over with beds of a broad-bladed turtle grass that swayed in the surge and current. In open places in the grass the sand was broken by protruding pen shells, like small gravestones, and scattered everywhere were limy skeletons of scallops, cockles and clams. While the cameras were being checked, I reached out of the cage to collect specimens. Then I remembered how Henri Bource's leg had been removed, and peered into the restless blue-gray murk. Overhead, at the silvering surface, hung the black oval of the Saori, where a pair of rotting sheep, heads down, did a grim dance with the ship's rise and fall.

Each night when the sun sinks beyond the Eyre Peninsula, the air turns cold. The nights are clear, and since Ben Ranford's cabin is quarters for five people besides himself (its spare berth and adjoining floor are nominally shared by Gimbel, Cody, myself and the Taylors), Gimbel and I sleep ordinarily on deck. In my own niche well forward by the anchor windlass, out of the way of night-time feet, I listen to the groan of the reef and the surf breaking on its weather shore and the wind of the universe spinning a full moon through blue-black skies. The Pacific gull adds its weird cry to the plaint of seals, and often the night voice sounds prehuman, a cry out of lost childhood or another world. Half asleep, I hear the calling of my name and listen fiercely, but there is nothing, only the black mainmast tossing, the black wind and whirling stars.

Saturday morning, January 24, and still no sign of shark. Gimbel is muttering about bad dreams. In one dream, he and Peter Lake had mutilated their legs with knives, then swum in the water to attract sharks. Their effort was in vain, and afterward they sat on the deck and compared their legs, which would have to be amputated. "We realized we'd made a mistake," Gimbel said, still worried by the dream; he looked up in surprise when everybody laughed, then laughed himself. In this time of stress, he is upset unduly by a cold sore at the corner of his mouth, which he chooses to see as an intimation of his own mortality.

The day is a good one, with light wind; the *Saori* will stay on here despite the absence of white sharks. But everyone is growing apprehensive. Even if sharks are numerous, it will be difficult to get good underwater film in the poor visibility of these cold turbid waters, especially if the whites avoid the air bubbles of the scuba tanks and cages. Dangerous Reef is a place notorious for the white shark ("The ab divers don't care for it," Bruce Farley says. "Took a quick look and decided they'd do better some place else"), yet no one has seen a fin. If only two years ago Rodney saw seven here at once, where are they now?

Rodney says that the sharks might be following the salmon schools that are moving through the gulf, and possibly their movements and feeding patterns have been affected by the past month of bad weather, which kept the water temperature well below the seasonal norm; though warmer than at Memory Cove, it is four degrees colder here than it was when Rodney came at this same time in 1968. In salt-water environments, a four-degree gradient makes a big difference – it can keep oysters from spawning, for example – and it has been noted with captive specimens as well as in shark attack statistics that a shark's interest in food rises markedly with an increase in temperature; in winter, captive sharks scarcely feed at all. The white shark is an exception to a general rule proposed by an Australian authority, V. M. Coppleson, that shark attacks around the world are rare in water colder than 72 degrees; nevertheless, its feeding may be inhibited by unseasonal cold water and unstable weather.

Or so we speculate. On the radio this morning, one of the tuna captains with a white shark charter party out of Port Lincoln had some disturbing news: in his last three trips to Dangerous Reef he had raised only one shy white shark, and it would not come near the boat. "Not like it used to be," the radio voice said. "They're being slowly killed out, I reckon, like everything else in the world." This man blamed the gill-netters for the disappearance of the sharks, and certainly commercial netting is a factor. Like other large predators of land and sea, the white shark will not survive long without the protection that it is unlikely to receive from man, and possibly the Australians are correct in the opinion generally held here that the species is nearing extinction. I am happy that our expedition has no plan to kill one except in self-defense.

There are recent reports of white sharks farther north in the gulf and to the west of Cape Catastrophe, and there is a feeling aboard that the *Saori* should pursue these sightings. Pursuit might relieve the strain of waiting and improve morale, but chasing works no better with fish than it does with anything else: better to pick one likely place and chum the hell out of it, day and night. By the time the slow *Saori* got the cages to the scene of any sighting, the shark might be twenty miles away, and even if a shark was present, there is no guarantee that water clarity would be adequate, or that the shark would approach the cage. Possibly inshore sharks have a hunting circuit, moving from point to point as wolves do, but more likely they move at random, taking prey as chance presents, and congregating now and then at likely grounds like Dangerous Reef. Instinctively, I agree with Captain Arno, who is relieving Captain Ben over this weekend. Arno is a wonderful bent old salt with white broken bare feet that never sunburn; offered grog, he smites the table, crying out fiercely, "I *will*!" Says Arno, "Sharks have a head and a tail, and they keep swimming. Nobody knows where the shark goes. I reckon they don't know where they are themselves."

Although these days are painful for Gimbel, they are almost as hard on Rodney Fox, who must choose the fishing grounds and baiting techniques that will bring the missing sharks. Rodney performs his duties with efficiency and style, but his casual air of cocky indifference is deceptive. If anything, he takes too much of the burden upon himself, and tends to construe the discussion of alternatives as implied criticism. For those aboard with a long interest in the sea and sharks, such discussions are fun and ease the strain, but giving a hearing to amateur opinions is hard on Rodney's nerves in a nervous time. Rarely does he permit the strain to show, but sometimes, muttering "Too many cooks . . . !" he lies face down on the deck, feigning deep sleep, and one day he actually took refuge in the hold, refusing to come out to eat his lunch.

Nevertheless, Rodney says, he has never worked with a nicer group of people. I feel the same, and so does everyone else; even Lipscomb and I, who often disagree, manage to disagree in a friendly manner. After a year and a half, this film crew is truly a unit, and its strength is mutual affection and acceptance: each man knows precisely what can be expected of the man beside him and demands no more, because those who fail in one respect have made it up over and over in others. As relationships have grown, the people have become more self-sufficient; even the hearts game, a loud nightly event aboard the *Terrier*, has given way to books and chess and backgammon.

There are other changes, in the crew's youngest members especially. A new confident Cody is so loose that he threatens to join the extroverts, while Lake has arrived at a new awareness in his dealings with others. One day on the deckhouse, watching for sharks, Peter said, "Remember when I wrote you that I didn't really care about this film? Well, that's all changed – I *do*."

By now everyone cares about the film, quite apart from his own investment in it, if only because everyone cares about Peter Gimbel, who has his life's work on the line. A great part of the suspense of waiting for "Big Whitey", as the near-mythical ruler of these silent seas has become known, is the knowledge that his failure to appear could be fatal to the film. Therefore the ship is quiet. Against these stark horizons, even the throb of hard rock music has a thin tinny ring.

More than once I went ashore and prowled the tide pools. I have spent hours of my life crouched beside tide pools, watching the slow surge of simple organisms still close to the first pulse of life on earth. On Dangerous Reef are gaudy giant limpets, and companies of blue, black and banded periwinkles, and the green snail and a brown cone and a very beautiful cream volute with zigzag stripings; also rockfish and the great fire-colored rock crabs that grow enormous in the deeps, and a heart-colored sea anemone, and a garden of hydroids, barnacles and algae. In every tide pool the seal pups played, and others lay on the warm rocks in a sleep so sound that they could be petted without awakening. When at last one did come to, it would stare for seconds in bare disbelief, then bleat in dismay and flop away at speed over the rocks.

In the white surge along the shore, the seals rolled endlessly, turning and twisting, whisking clean out of the water in swift chases, or ranging along, the sleek sunshined dark-eyed heads held high out of the sea. A small surge would lift them out onto the granite where, groaning, they dozed on the hot rocks in rows. The old bulls, though graceful in the water, were less playful; they stationed themselves on underwater ledges like old mighty sentinels and let the white foam wash around them. Onshore, competitors were driven off in heaving neck fights that were mostly shoving contests; the animals swayed their heavy heads and necks in the way of bears, to which, among land mammals, they are most closely related. Sea lions are agile on the land, and a golden-maned bull protecting a cow and a new pup drove me up onto high ground. One cow was raked drastically on her hind end and right hind flipper by the parallel black lines of an old shark bite, and it was noticeable that the young never left the shallows and that even the adults kept close to the reef edges when not off at sea.

At noon today the *Sea Raider* brought word that an eleven-foot white of thirteen hundred pounds had been hooked at Cape Donington, where the *Saori* had anchored two nights before. Psychologically this news was painful, but the water clarity at Cape Donington is awful and we could not have worked there. And at least it was proof that the species was not extinct.

In a letter to a friend this morning, Valerie wrote that no shark had been seen, but that she expected a twelve- or thirteen-footer to turn up at about 2.00. At 2.20, Peter Lake and Ian McKechnie saw a fin in the slick, some fifty yards behind the ship: the spell was broken. We dragged on diving suits and went on watch, but the fin had sunk from view in the still sea. A half hour passed, and more. Then, perhaps ten feet down off the port beam, a fleeting brown shadow brought the sea to life.

Suspended from a buoy, a salmon was floated out behind the boat to lure the shark closer. Once it had fed at the side of the boat, it would be less cautious; then, perhaps, the engine could be started and the cages swung over the side without scaring it away. But an hour passed before the shark was seen again. This time a glinting rusty back parted the surface, tail and dorsal high out of the water as the shark made its turn into the bait; there was the great wavering blade exactly as Al Giddings had described it, and the thrash of water as the shark took the salmon, two hours to the minute after the first sighting. Stan Waterman cried, "Holy sweet Jesus!", a very strong epithet for this mild-spoken man; he was amazed by the mass of shark that had been raised clear of the water. Even the Australians were excited, try as they would to appear calm. "Makes other sharks look like little frisky pups, doesn't it?" cried Valerie with pride. Then it was gone again. Along the reef, a hundred yards away, the sea lions were playing tag, their sleek heavy bodies squirting clean out of the water and parting the surface again without a splash, and a string of cormorant, oblivious, came beating in out of the northern blue.

Gimbel, annoyed that he had missed the shark, was running from the bow; he did not have long to wait. From the deckhouse roof, I could see the shadow rising toward the bait. "There he is," I said, and Rodney yanked at the shred of salmon, trying to bring the shark in closer to the ship. Lipscomb, beside me, was already shooting when the great fish breached, spun the sea awash and lunged after the skipping salmon tail; we stared into its white oncoming mouth. "My *God*!" Gimbel shouted, astounded by the sight of his first white shark. The conical snout and the terrible shearing teeth and the dark eye like a hole were all in sight, raised clear out of the water. Under the stern, with an audible *whush*, the shark took a last snap at the bait, then wheeled away; sounding, it sent the skiff spinning with a terrific whack of its great tail, an ominous boom that could have been heard a half mile away.

For a split second there was silence, and then Lipscomb gave a mighty whoop of joy. "I *got* it!" he yelled. "Goddamn it, I *got* it!" There was a bedlam of relief, then another silence. "Might knock that cage about a bit," Rodney said finally, hauling in the shred of fish; he was thinking of the baits that would be suspended in the cage to bring the shark close to the cameras. Gimbel, still staring at the faceless water, only nodded.

Just after 5.00 the shark reappeared. The late sun glistened on its dorsal as it cut back and forth across the surface, worrying a dead fish from the line. There was none of the sinuous effect of lesser sharks; the tail strokes were stiff and short like those of swordfish, giant tuna and other swift deep-sea swimmers. This creature was much bigger than the big oceanic sharks off Durban, but for a white shark it was not enormous. Estimates of its length varied from eleven feet six inches ("Ron always plays it safe and underestimates," said Valerie) to fourteen feet ( Peter Gimbel: "I saw it alongside that skiff and I'm certain it was at *least* as long – I'm *certain* of it!"), but much more impressive than the length was the mass of it, and the speed and power. "It doesn't matter *what* size the bastards are," Rodney said. "A white shark over six feet long is bloody dangerous."

The day was late. In the westering sun, a hard light of late afternoon silvered the water rushing through the reef, and nearer, the blue facets of the sea sparkled in cascades of tiny stars. More out of frustration than good sense, the choice between trying to film the shark immediately and trying to lure it to the baits alongside, in the hope of keeping it nearby overnight, was resolved in favor of immediate action. The motor was started up and the cages swung over the side, and the cameramen disappeared beneath the surface. But the great shark had retreated, and did not return.

By dark the wind exceeded 25 knots, and went quickly to 30, 40, and finally, toward 1.00 in the morning, to 50 or better – a whole gale. On deck, I lay sleepless, rising every little while to check the position of the light on Dangerous Reef. The Reef is too low to make a windbreak, and even close under the lee, the *Saori* tossed and heaved under heavy strain. But Captain Ben, who knew exactly what his ship would do, slept soundly below. Toward 3.00 the wind moderated, backing around to the southeast, where it held till daybreak.

This morning the wind has died to a fair breeze. Waiting, we sit peacefully in the Sunday sun. The boat captains hand-line for Tommy-rough, a delicious small silver relative of the Australian "salmon". Others tinker with equipment, play chess and backgammon, write letters and read. Peter Lake has put a rock tape on the sound machine, and on the roof of the pilothouse, overlooking the oil slick, I write these notes while listening to The Band. Onshore, for Jim Lipscomb's camera, Valerie in lavender is baby-talking with baby seals, and I hope that most if not all of this sequence will die on the cutting-room floor. Unless it points up the days of waiting, such material has no place in the climax of the film; it will soften the starkness of this remote reef as well as the suspense surrounding the imminence of the white shark. Stan and Valerie, with a background of lecture films and a taste for amateur theatrics, share Jim's appetite for "human-interest stuff", which might yet reduce this film to the first million-dollar home movie.

Toward dark another shark appeared, a smaller one, much bolder. Relentlessly it circled the ship, not ten feet from the hull. On one pass it took the buoyed tuna at a single gulp.

Since it passed alongside, the size of this shark could be closely estimated: all hands agreed that it was between nine feet and ten. But if this was accurate, the shark yesterday had been larger than was thought. Rodney now said that it was over twelve, Valerie between thirteen and fourteen, and Gimbel thought that it might have been sixteen feet: "I thought so *yesterday*," he said, "but I felt foolish, with everyone else saying twelve." I thought thirteen feet seemed a conservative minimum. In any case, it had twice the mass of tonight's shark, which was plenty big enough. As it slid along the hull, the thick lateral keel on its caudal peduncle was clearly visible; the merest twitch of that strong tail kept it in motion. Underwater lights were lit to see it better, but this may have been a mistake; it vanished, and did not return the following day.

On January 26 the *Saori* returned to port for water and supplies. There it was learned that four boats, fishing all weekend, had landed between them the solitary shark that we had heard about on Saturday. The *Saori* could easily have hooked two, but what she was here for was going to be much more difficult. Meanwhile, a sighting of white sharks had been reported by divers working Fisheries Bay, west of Cape Catastrophe on the ocean coast, where three whites and a number of bronze whalers had been seen schooling behind the surf; the bronze whaler, which may be the ubiquitous bull shark, *C. leucas*, is the chief suspect in most shark attacks on Australia's east coast.

On the chance that the shark school was still present, we drove out to the coast across the parched hills of the sheep country. Over high, wind-burnt fields a lovely paroquet, the galah, pearl-gray and rose, flew in weightless flocks out of the wheat; other paroquets, turquoise and black and gold, crossed from a scrub of gum trees and melaleuca to a grove of she-oak, the local name for a form of casuarina. Along the way were strange birds and trees in an odd landscape of wind-worn hills that descended again to the sea-misted shore. From the sea cliffs four or five whalers were in sight, like brown ripples in the pale-green windy water, but the white sharks had gone.

AT DAYBREAK ON Wednesday the *Saori* sailed for the Gambier Islands, on the Antarctic horizon south of the mouth of Spencer Gulf. A big ocean swell rose out of the southwest, from the far reaches of the roaring forties, but there was a lee of sorts east of Wedge Island. The Gambiers are remote and no gill netting is done there, and white sharks had been seen often in the past; occasionally the sharks would seize a horse when the animals raised here in other days were swum out to the ships. Now the old farm was a sheep station, visited infrequently by man. With Ron, Valerie and Stan, I went ashore, exploring. Gaunt black machinery, stranded by disuse, looked out to sea from the dry golden hills, and the sheep, many of them dead, had brought a plague of flies; only at the island's crest, in the southwest wind, could one be free of them.

Wedge Island is a beautiful silent place, a great monument like a pyramid in the Southern Ocean. That night, white-faced storm petrels fluttered like moths at the masthead light. Some fell to the deck, and I put them in a box; once the deck lights were out, they flitted off toward the island. These hardy little birds come in off the windy wastes of sea just once a year to nest in burrows in the cliffs.

Overhead, shined by the wind, the austral sky was luminous. With the stem of his pipe, Ben Ranford pointed at the universe: "Canis Major," he pronounced with satisfaction. "The brightest star in all the heavens." In World War II Ben was captain of a destroyer in the Australian Navy, and is still the compleat seaman, clumping here and there about his ship in white coveralls and big black shoes without one wasted motion; he could have stripped the *Saori* from stem to stern and reassembled her in the dark. No man could do his job better than he, and yet Ben knew that this ship might be his last.

* * *

At dawn the day was already hot and still, the baits untouched, the ocean empty. Only a solitary eagle, white head shimmering in the rising sun, flapped and sailed over the sea, bound for the outermost islands.

Two weeks had passed, and there was no underwater footage, and running from place to place was not the answer. A decision was made to increase the volume of bait and chum and concentrate it at Dangerous Reef. The two sharks raised there were the only two that had been seen, the resident sea lions were an asset, and the Reef was only three hours from the abattoirs and fish companies at Port Lincoln. The ship sailed north again into Spencer Gulf, rounding the west end of the Reef and anchoring off its northern shore at noon; a southwest blow was expected that afternoon, backing around to the southeast by evening.

White shark number three came after dark on January 27, seizing the floating bait with a heavy thrash that brought a bellow of excitement from Gimbel, working on deck. No sooner had a light been rigged than the fish reappeared, making a slow turn at the perimeter of green night water. Then it rifled straight and fast for a carcass hanging at the ship's side, which it gobbled at and shook apart, oblivious of the lights and shouting men. Though not enormous, this aggressive brute was the one we wanted; by the look of it, it would not be deterred by cages or anything else. Then it was gone, and a cuttlefish rippled in the eerie light, and the sea thickened with a bloom of red crustaceans.

All baits were hauled in but a small flayed sheep, left out to stay the shark until the morning. At dawn, the unraveled bait line lay on deck. Taking the sheep, the shark had put such strain upon the line that, parting, it had snapped back clean out of the water. But there was no sign of the shark, and it never returned.

That morning the *Sea Raider* came out from Port Lincoln with big drums full of butchered horse; the quarters hung from the stern of the *Saori*, which was reeking like a charnel house. Buckets of horse blood, whale oil and a foul chum of ground tuna guts made a broad slick that spread northeast toward Spilsby Island. The cages, cameras lashed to their floors, were already overboard, floating astern. The sky was somber, with high mackerel clouds and a bank of ocean grays creeping up out of the south, and a hard wind; petrels dipped and fluttered in the wake. The ship was silent.

Vodka in hand, Gimbel came and went, glaring astounded at the empty slick that spread majestically to the horizon. About 5.30 I forsook my post on the deckhouse roof and went below. Peter was lying in a berth, face tight. I said, "I'm taking a shower even though there's still light enough to shoot; there'll be a shark here before I'm finished." He laughed politely. I had just returned to the cabin, still half dry, wrapped in a towel, when a voice yelled "Shark!" down the companionway.

By the time we reached the deck, bound for the wet-suits, the sun had parted the clouds; with luck, there would be underwater light for at least an hour. Already a second shark had joined the first, and both were big. I went into the sea with Peter, and Stan and Ron soon joined us in the other cage. Almost immediately a great pale shape took form in the blue mist.

The bolder of the sharks, perhaps twelve feet long, was a heavy male, identifiable by paired claspers at the vent; a second male, slightly smaller, stayed in the background. The first shark had vivid scars about the head and an oval scar under the dorsal, and in the molten water of late afternoon it was a creature very different from the one seen from the surface. The hard rust of its hide had dissolved in pale transparent tones that shimmered in the ocean light on its satin skin. From the dorsal fin an evanescent bronze shaded down to luminous dark metallic gray along the lateral line, a color as delicate as that bronze tint on a mushroom which points up the whiteness of the flesh beneath. From snout to keel, the underside was a deathly white, all but the black undertips of the broad pectorals.

The shark passed slowly, first the slack jaw with the triangular splayed teeth, then the dark eye, impenetrable and empty as the eye of God, next the gill slits like knife slashes in paper, then the pale slab of the flank, aflow with silver ripplings of light, and finally the thick short twitch of its hard tail. Its aspect was less savage than implacable, a silent thing of merciless serenity.

Only when the light had dimmed did the smaller shark drift in from the blue shadows, but never did it come to the hanging baits. The larger shark barged past the cages and banged against the hull to swipe and gulp at the chunks of meat; on the way out, it repeatedly bit the propeller of the outboard, swallowing the whole shaft and shaking the motor.

Then it would swing and glide straight in again, its broad pectorals, like a manta's wings, held in an upward curve. Gills rippling, it would swerve enough to miss the cage, and once the smiling head had passed I could reach out and take hold of the rubber pectoral, or trail my fingers down the length of cold dead flank, as if stroking a corpse: the skin felt as smooth as the skin of a swordfish or tuna. Then the pale apparition sank under the copper-red hull of the *Saori* and vanished in the gloom, only to reappear from another angle, relentless, moving always at the same deceptive speed, mouth gasping as in thirst. This time it came straight to the cage and seized one of the flotation cylinders of the cage roof; there came a nasty screeching sound, like the grating of fingernails on slate, before the shark turned off, shaking its head.

The sharks off Durban had probed the cages and scraped past, but never once, in hundreds of encounters, did one attack them open-mouthed. The white sharks were to attack the cages over and over. This first one arched its back, gills wrinkling, coming on mouth wide; fortunately it came at cruising speed and struck the least vulnerable part of the cage. The silver tanks, awash at the surface, may have resembled crippled fish, for they were hit far more often than anything else. When their teeth struck metal, the sharks usually turned away, but often the bite was hard enough to break the teeth out. Sometimes as it approached the cage, one would flare its mouth wide, then close it again, in what looked very much like the threat display of higher animals.

To escape the rough chop at the surface, the cage descended to fifteen feet, where Gimbel opened the roof hatch and climbed partway out to film; he was driven back each time. At one point, falling back in haste, Peter got his tank hung up on the hatch, and was still partly exposed when the shark passed overhead, a black shade in the golden ether made by the sinking sun. From below, the brute's girth was dramatically apparent; it blotted out the light.

The shark paid the cages such close attention that Gimbel burned up a ten-minute magazine in fifteen minutes. When he went to the surface to reload, Valerie Taylor and Peter Lake took over the cage. "Listen!" Gimbel yelled at them, still excited. "Now watch it! They're nothing like those Durban sharks, so don't take chances!" Then Stan came out of the second cage, and by the time he was reloaded, Ron was ready to come out; this gave me a chance to go down a second time.

For a while the atmosphere was quiet as both sharks kept their distance from the ship; they came and went like spirits in the mist. But emergencies are usually sudden, and now there came a series of near emergencies. First the bigger shark, mouth open, ran afoul of one of the lines; the length of rope slipped past the teeth and hung in the corners of its mouth, trailing back like reins. So many lines were crisscrossed in the water – skiff lines, bait lines, hydrophone cable and tethers to keep the cages near the bait – that at first one could not tell what was going to happen, and I felt a clutch of fear. Swimming away, the shark was shaking its head in irritation, and then I saw that the line was the tether of the other cage, where Gimbel had been joined by Peter Lake. The line was very nearly taut when the shark shook free. Lake was using a camera with a 180-degree "fish-eye" lens, and was getting remarkable shots, but the close call rattled him considerably. At the surface, he yelled all the obscenities. "To hell with that shit," he concluded. "I'm going below to hide under my berth!" But Lake's trials were not over. A few days later, when the *Saori* returned to Dangerous Reef for continuity shots and supporting footage, a shark, tangled in a bait line, bent the whole cage with its slow thrashing; it actually *stretched* five of the bars, shaking the whole cage like a dice cup before Lake could get his leg knife out and cut it free. At the surface, he had difficulty joking: "When I saw those bars starting to go I felt like I had jumped at twelve thousand feet with my parachute eaten by rats."

Often the larger shark would appear from below, its ragged smile rising straight up past the cage; already its head was scarred with streaks of red lead from the *Saori*'s hull. On one of these ascents it seized a piece of meat hung from the taffrail just as the current swung the cage in toward the ship, so that the whole expanse of its ghostly belly, racked by spasms of huge gulping, was perpendicular against the bars. I scratched the belly with a kind of morbid sympathy, but at that instant we were jarred by a thrash of the tail; the cage had pinned the shark upright against the rudder of the *Saori*. While Waterman filmed at point-blank range, it lashed the water white. "I wasn't really worried about you guys," Gimbel said later. "I just knew it would knock hell out of you." The cage was swiftly heaved aside, and the shark glided for the bottom with that ineffable silent calm, moving no more rapidly than before. Except for size, it is often difficult to estimate shark age, and watching it go, it was easy to believe that this beast might swim for centuries.

I turned to congratulate Waterman on the greatest footage of a feeding white shark ever taken, but bald eyes rolled in woe behind his mask, and he made a throat-slitting gesture with his finger and smote his rubber brow, then shook his fist at his camera, which had jammed. Gimbel got the sequence from the other cage, thirty feet away, and Lipscomb caught one angle of it from the surface, but Stan was inconsolable.

Gimbel was still trying to film from the roof hatch, and now he ducked down neatly at a shark's approach, only to find himself staring straight into its face. The main cage door had opened outward, and the shark was so near that he could not reach out to close it. Badly frightened, he feinted with his camera at the shark, which cruised on past, oblivious.

Between bites the sharks patrolled the cages, the *Saori* and the skiff, biting indiscriminately; there was no sense of viciousness or savagery in what they did, but something worse, an implacable need. They bit the skiff and they bit the cages, and one pushed past the meat to bite the propeller of the *Saori*; it was as if they smelled the food but could not distinguish it by sight, and therefore attacked everything in the vicinity. Often they mouthed the cage metal with such violence that teeth went spinning from their jaws. One such tooth found on the bottom had its serrated edge scraped smooth. It seemed to me that here was the explanation for the reports of white shark attacks on boats; they do not attack boats, they attack *anything*.

When I left the water, there was a slight delay in getting the skiff alongside, and Rodney warned me not to loiter on top of the cage. "They've been climbing all over it!" he called. At one point Valerie, having handed up her exhausted tank, had to retreat into the cage, holding her breath as a shark thrashed across its roof over and over.

We had entered the water about 6.00, and the last diver left it at 7.30, by which time every one of us was shaking hard with cold. In the skiff, transferring from the cages to the ship, people were shouting. The excitement far exceeded any I had seen in the footage of the greatest day off Durban; as Gimbel said, "Christ, man! These sharks are just a hell of a lot more exciting!"

The next morning, a sparkling wild day, the two sharks were still with us, and they had been joined by a third still larger. Even Ron estimated the new shark at fourteen feet, and Gimbel one or two feet more; it was the biggest man-eating shark that anyone aboard had ever seen. Surging out of the sea to fasten on a horse shank hung from a davit, it stood upright beside the ship, head and gills clear of the water, tail vibrating, the glistening triangles of its teeth red-rimmed with blood. In the effort of shearing, the black eye went white as the eyeball was rolled inward; then the whole horse quarter disappeared in a scarlet billow. "I've watched sharks all my life," Ben Ranford said, but I've never seen anything as terrifying as that." Plainly no shark victim with the misfortune to get hold of a raft or boat would ever survive the shaking of that head. Last night in the galley, Ron had suggested to Peter that swimming with one white might be possible, and Peter agreed. But this morning there were three, and the visibility was so limited that one could never tell where or when the other two might appear. The talk of swimming in the open water ended, and a good thing, too. In its seeming contempt for the great white shark, such a dangerous stunt could only make an anticlimax of the film's climax.

The cage will sink a foot or so beneath the surface under a man's weight – a situation to be avoided in the presence of white sharks – and the next morning, entering it, I performed with ease what I had heretofore done clumsily, flipping directly out of the skiff and down through the narrow roof hatch head first. Even before I straightened up, the largest of the sharks loomed alongside, filling the blue silence with its smile. I felt naked in my flimsy cell until Stan joined me. This shark was two or three feet longer than the next in size, but it looked half again as big, between eighteen hundred pounds and a fat ton. In white sharks over ten feet long, the increase in girth and weight per foot of length is massive; the white shark that I saw dead at Montauk, only two or three feet longer than this one, had weighed at least twice as much.

The new shark was fearless, crashing past skiff and cage alike to reach the meat, and often attacking both on the way out. Like its companions, which scooted aside when it came close, it attacked the flotation tanks over and over, refusing to learn that they were not edible. Even the smallest shark came in to sample the flotation tanks when the others were not around. I had seen one of its companions chase it, so probably its shyness had little to do with the *Saori*: unlike the sharks in the Indian Ocean the whites gave one another a wide berth. Occasionally one would go for the air tank in the corner, bumping the whole cage through the water with its snout, and once one struck the naked bars when I waved a dead salmon as it approached. Clumsily it missed the proffered fish, glancing off the bars as I yanked my arm back. Had the sharks attacked the bars, they would have splayed them. "He could bite that cage to bits if he wanted to," Valerie had said of yesterday's shark, and got no argument; for the big shark today, the destruction of the cage would be the work of moments. From below, we watched it wrestle free an enormous slab of horse, two hundred pounds or more; as it gobbled and shook, its great pale body quaked, the tail shuddering with the effort of keeping its head high out of the water. Then, back arched, it dove with its prize toward the bottom, its mouth trailing bubbles from the air gulped down with its last bite. Only one pilotfish was ever seen at Dangerous Reef; we wondered if the white shark's relentless pace made it difficult for a small fish to keep up.

Numbers of fish had come to the debris exploded into the water by the feeding, and the windstorm of the night before had stirred pale algae from the bottom. Visibility was poor, yet the sharks worked so close to the cages that the morning's filming was even better than the day before, and the cameramen worked from nine until one-thirty. By then, the ten months of suspense were over.

We were scarcely out of the water when the wind freshened, with the threat of rain. The cages were taken aboard and battened down while a party went ashore to film the *Saori* from the Reef. Then in a cold twilight, drinking rum in the galley-fo'c's'le, we rolled downwind across Spencer Gulf, bound for Port Lincoln. Though the sea was rough, the fo'c's'le was warm and bright, filled with rock music. Valerie saw to it that the supper was cooked properly, and wine soon banished the slightest doubt that we all liked one another very much. "Is there anything more splendid," Waterman cried, "than the fellowship of good shipmates in the fo'c's'le after a bracing day before the mast?" After three weeks in the fo'c's'le, Stan had embraced the nineteenth century with all his heart.

Peter Gimbel, sweetly drunk, swung back and forth from fits of shouting to a kind of stunned suffused relief and quiet happiness. He looked ten years younger. What was surely the most exciting film ever taken underwater had been obtained without serious injury to anybody. The triumph was a vindication of his own faith in himself, and because he had earned it the hard way and deserved it, it was a pleasure simply to sit and drink and watch the rare joy in his face.

At the end of that week all the Americans returned home but myself and Cody. Stuart went into the Outback to try opal mining with Ian McKechnie, and I flew westward to East Africa. A month later, when I reached New York, Peter told me that the white shark sequence was beyond all expectations, that the film studio was ecstatic, and that a financial success now seemed assured. How sad, I said, that his father wasn't here to see it. He grinned, shaking his head. "It is," he said. "He would have been delighted."

Already Peter was concerned about where he would go from here. Meanwhile, he had planned a violent dieting which he didn't need, and when asked why, he shrugged. "I just want to see if I can get down to a hundred seventy," he said. Perhaps I read too much into that diet, but it bothered me: the search for the great white shark was at an end, but the search was not. I recalled a passage in the letter Peter had written after the thirty-hour marathon off Durban, and when I got home I dug it out.

"I felt none of the dazed sense of awe," he wrote, "that had filled me ten days before during our first night dive. I remember wondering sadly how it could be that a sight this incredible could have lost its shattering impact so quickly for me – why it should be that the sights and sensations should have to accelerate so hellishly simply to hold their own with my adaptation to them . . . Only a week or so after having come out of the water one night to say over and over, 'No four people in all the world have ever laid eyes on a scene so wild and infernal as that,' I wasn't even particularly excited . . ."